Springer Tracts in Modern Physics 113

Springer Tracts in Modern Physics

87 **Point Defects in Metals II:** Dynamical Properties and Diffusion Controlled Reactions With contributions by P. H. Dederichs, K. Schroeder, R. Zeller

88 **Excitation of Plasmons and Interband Transitions by Electrons** By H. Raether

89 Giant Resonance Phenomena in **Intermediate-Energy Nuclear Reactions** By F. Cannata, H. Überall

90* **Jets of Hadrons** By W. Hofmann

91 **Structural Studies of Surfaces** With contributions by K. Heinz, K. Müller, T. Engel, and K. H. Rieder

92 **Single-Particle Rotations in Molecular Crystals** By W. Press

93 **Coherent Inelastic Neutron Scattering in Lattice Dynamics** By B. Dorner

94 **Exciton Dynamics in Molecular Crystals and Aggregates** With contributions by V. M. Kenkre and P. Reineker

95 **Projection Operator Techniques in Nonequilibrium Statistical Mechanics** By H. Grabert

96 **Hyperfine Structure in 4d- and 5d-Shell Atoms** By S. Büttgenbach

97 **Elements of Flow and Diffusion Processes in Separation Nozzles** By W. Ehrfeld

98 **Narrow-Gap Semiconductors** With contributions by R. Dornhaus, G. Nimtz, and B. Schlicht

99 **Dynamical Properties of IV–VI Compounds** With contributions by H. Bilz, A. Bussmann-Holder, W. Jantsch, and P. Vogl

100* **Quarks and Nuclear Forces** Edited by D. C. Fries and B. Zeitnitz

101 **Neutron Scattering and Muon Spin Rotation** With contributions by R. E. Lechner, D. Richter, and C. Riekel

102 **Theory of Jets in Electron-Positron Annihilation** By G. Kramer

103 **Rare Gas Solids** With contributions by H. Coufal, E. Lüscher, H. Micklitz, and R. E. Norberg

104 **Surface Enhanced Raman Vibrational Studies at Solid/Gas Interfaces** By I. Pockrand

105 **Two-Photon Physics at e^+e^- Storage Rings** By H. Kolanoski

106 **Polarized Electrons at Surfaces** By J. Kirschner

107 **Electronic Excitations in Condensed Rare Gases** By N. Schwentner, E.-E. Koch, and J. Jortner

108 **Particles and Detectors** Festschrift for Jack Steinberger Edited by K. Kleinknecht and T. D. Lee

109 **Metal Optics Near the Plasma Frequency** By F. Forstmann and R. R. Gerhardts

110* **Electrodynamics of the Semiconductor Band Edge** By A. Stahl and I. Balslev

111 **Surface Plasmons** on Smooth and Rough Surfaces and on Gratings By H. Raether

112 **Recent Experimental Tests of the Standard Theory of Electroweak Interactions** By C. Kiesling

113 **Radiative Transfer in Nontransparent, Dispersed Media** By H. Reiss

---

* denotes a volume which contains a Classified Index starting from Volume 36

Harald Reiss

Radiative Transfer in Nontransparent, Dispersed Media

With 72 Figures

Springer-Verlag
Berlin Heidelberg New York
London Paris Tokyo

Privatdozent Dr. Harald Reiss
Asea Brown Boveri, Corporate Research Heidelberg, Eppelheimer Straße 82
D-6900 Heidelberg, Fed. Rep. of Germany

Manuscripts for publication should be addressed to:
Gerhard Höhler
Institut für Theoretische Kernphysik der Universität Karlsruhe
Postfach 6980, D-7500 Karlsruhe 1, Fed. Rep. of Germany

Proofs and all correspondence concerning papers in the process of publication should be addressed to:
Ernst A. Niekisch
Haubourdinstraße 6, D-5170 Jülich 1, Fed. Rep. of Germany

ISBN 3-540-18608-5 Springer-Verlag Berlin Heidelberg New York
ISBN 0-387-18608-5 Springer-Verlag New York Berlin Heidelberg

Library of Congress Cataloging-in-Publication Data. Reiss, Harald, 1941–. Radiative transfer in nontransparent, dispersed media/Harald Reiss. p. cm. – (Springer tracts in modern physics; 113) Bibliography: p. Includes index. 1. Radiative transfer. I. Title. II. Series. QC320.R45 1988 536'.33–dc 19 87-32253

The text was word-processed using PS™ software and was printed with a Toshiba P321

Printing and binding: Brühlsche Universitätsdruckerei, 6300 Giessen
2153/3150-543210

Preface

This book is addressed to physicists who are interested in an introduction to the theoretical and experimental methods that are used to determine heat flow components in powders, fibres, sand, dust and other dispersed media. The title of this book stresses the key role occupied by radiative transfer when the total heat flow is separated into the various transfer modes.

Various presentations of the problem of radiative transfer already exist. S.Chandrasekhar's famous book "Radiative Transfer" [1950] is the definitive work in the field (it has also become so well known that V. Kourganoff in his book "Basic Methods in Transfer Problems" [1952] referred to it simply as "R.T."). The contributions of H.C. Hottel and A.F. Sarofim, E.M. Sparrow and R.D. Cess, R. Siegel and J.R. Howell, to mention only a few, have also described the radiative transfer problem on an excellent level. However, Chandrasekhar's treatise is a purely theoretical textbook, and although the other authors have treated the problem of radiative transfer primarily from an engineering standpoint, only minor attention is paid to the confirmation of theoretical predictions on radiative flow by experimental, especially calorimetrical, means. The radiative transfer problem is perhaps the most exciting of the various heat transfer mechanisms in dispersed media; nevertheless it is only *one* heat transfer mode. Because of energy conservation, the other heat transfer modes influence radiative transfer, and vice versa. An understanding of the radiative transfer cannot be decoupled from an understanding of the nonradiative transfer mechanisms. Accordingly, this book contains additional material which is not found in other radiative transfer volumes.

Chapter 1 gives an introduction to well-known fundamental heat transfer problems such as the coupling of heat transfer modes by temperature profiles, and the definition of thermal conductivity as a unique quantitiy, among others. Chapter 2 deals with temperature dependencies of nonradiative heat transfer modes (gaseous and contact heat flow) in dispersed media, and it gives an introduction to the calculation of extinction coefficients using Mie theory, and to the problem of dependent scattering. It contains also a list of references for refractive indices in the infrared spectral region. Chapter 3 reviews theoretical models to describe radiative transfer in absorbing and emitting, scattering and heat conducting media. Because the comparison of theoretical predictions with calorimetric measurements is one of the most important subjects of this book, application of the diffusion model for radiative transfer is considered in detail. Chapters 4 and 5 are devoted to a comparison of calorimetric and spectroscopic measurements with theoretical predictions on extinction coefficients of dispersed media. Finally, Chap. 6 describes how optimum radiation extinction can be achieved.

I would like first to thank Prof. J. Fricke, Drs. R. Caps and D. Büttner and the whole thermophysics group of Würzburg University for their cooperation since 1980. Without their numerous valuable contributions, this book would probably not have been written. Thanks are also due to my colleagues Drs. B. Ziegenbein and H. Birnbreier for experimental collaboration and for stimulating discussions, and to Dr. F. Gross, former head of the BBC research laboratory, for his continuous interest in and support of this work. I am very grateful to Prof. P.G. Klemens, University of Storrs, CT, Drs. P. Morrell and B. Geary, University of Manchester, Institute of Science and Technology, and to M.F. Daly, B.Sc., University of Strathclyde, Glasgow, for reading and correcting the manuscript. My thanks are also extended to Drs. A. Lahee and H. Lotsch, Springer-Verlag, Heidelberg, for their friendly cooperation, and to Mrs. Frasure, University of Karlsruhe, and to Ms. B. Staubitz, Springer-Verlag, Heidelberg, for carefully typing the manuscript. Any errors appearing in this book are solely mine, however.

Heidelberg, October 1987 *Harald Reiss*

Contents

1. General Introduction into the Determination of Heat Flow Components

All real substances absorb electromagnetic radiation in a certain range of wavelengths. Consequently, part of the incident radiation is transformed into thermal radiation. If the medium is a thermal conductor and if a temperature sink exists, incident radiation transforms into conductive heat flow by conservation of energy. A similar consideration applies to the creation of convective heat flow. With very few exceptions, investigation of radiative transfer cannot be decoupled from an appropriate understanding of *total* heat flow propagation.

Although this work considers primarily thermal incident radiation, the methods of determination of heat flow components in a conductive medium, as discussed below, apply equally well to arbitrary wavelength regions of incident, e.g. visible, radiation. However, interactions between constituents of a medium and incident radiation will be considered only from the viewpoint of Maxwell's equations, i.e. from a classical continuum aspect. Consequently, the absorption and scattering properties of a medium are characterized exclusively by *macroscopic* variables and boundary conditions, i.e. complex refractive indices, $m = n-i\cdot k$, and particle geometry. To give an example, absorption will be treated only by the introduction of a nonvanishing imaginary part k of m, and no explicit reference is made to electronic transitions, creation of annihilation of particles or particle holes, excitation of vibrational states or other microscopic absorption processes.

This work summarizes calorimetric, spectroscopic and numerical methods that have been applied in the literature to separate heat flow components and analyze temperature profiles in dispersed media. Radiative transfer will be shown to play a key role in providing an insight into the interaction of heat transfer modes.

1.1 Why is Determination of Heat Flow Components Important?

Before we restrict our discussion, for the reasons outlined in Sect.1.2, to the propagation of radiation in nontransparent[1] media, let us briefly illustrate the importance of a determination of heat flow components in arbitrary (gaseous, liquid or solid) substances.

All gases are transparent at certain wavelengths. Thermal conductivity of gases can be specified only if all radiative contributions to the total, i.e. calorimetrically measured, heat flow are eliminated. The same problem applies to transparent liquids. The specific structure of a temperature profile observed in a transparent or nontransparent medium can be explained only if the contributions of all existing heat transfer modes are known. In turn, a temperature profile can be considered as an indicator that tells us which heat transfer modes are dominating. Approximately linear temperature profiles usually indicate large conductive heat flow and large radiation extinction. In the interior of stars, a large temperature gradient due to strong absorption in the atmosphere causes convection, which is frequently the dominant form of energy transfer. A curved temperature profile in a planar homogeneous medium, e.g. in a glass melt, indicates in most cases strong radiative flow in certain spectral regions. The curved profile in a red-hot glass lens manifests itself in the progress of solidification from boundaries to the interior. Obviously, determination of heat flow components is of particular

[1] A quantitative definition of nontransparency will be given by means of (1.3-5). At this stage, it suffices to define nontransparency as follows: Let us place a tunable laser in front of a nontransparent, nonconductive medium. On the other side of the medium, a radiation detector will be positioned exactly on the axis of the laser beam. Since it is assumed that the medium is nontransparent, the detector cannot see the original beam. As a consequence, the detector can neither distinguish between monochromatic radiation emitted by the laser at different intensities nor can it separate radiation of constant intensity that is emitted at different wavelengths. Note that this definition of nontransparency is related to *extinction* (of a beam). On the other hand, if the detector sees some radiation, this is not necessarily the consequence of transparency. The medium though nontransparent could be translucent, because of scattering properties that might dominate the modes of radiation-particle interactions.

importance in thermal insulators. Optimization of insulating properties will be successful only if all heat flow components can be reduced simultaneously. A great part of the experimental work reviewed below has dealt with material for thermal insulation applications. Although thermal insulation is usually associated with amorphous powders, with glass or ceramic fibres or with organic foams, thermal insulations do *not* constitute a special class of materials. Experience shows that a surprisingly large variety of solid substances (natural or artificial) can be used for insulation purposes provided they are available in highly porous states.[2]

The spectroscopic and calorimetric investigations reported here are as a rule directed towards the pure, original, dispersed material, that is, the results reviewed enjoy a much broader validity than thermal insulation species could occupy. These materials are readily available for laboratory experiments, in contrast to stellar dust or samples from the Earth's interior. Experiments using these materials are thus a sensitive method of corroborating theoretical heat flow models. Also, the wide spectrum of chemical composition, particle geometry, electrical conductivity and optical properties, and the fascinating structure of particle aggregates makes dispersed solid materials most attractive for research into their thermophysical properties.

Furthermore, it is important to analyze radiative flow in chemical or nuclear reactors, ablative layers, plasmas[3], geological structures, e.g. for understanding the temperature profile in the Earth's crust.

Unfortunately, there is no experimental method that allows a direct determination of a single heat flow component (a calorimetric measurement always yields total heat flow). All attempts described

---

[2] From a biogenetic point of view, clothing is also thermal insulation. The same applies to a nomadic tent or to an igloo. Nature efficiently uses low thermal conductivity of air for insulation purposes: see the photograph of a bird imprisoned in a climatic chamber [Reiss 1985b].

[3] It is well known that a plasma, e.g. the ionosphere, is nontransparent if the frequency ω of incoming radiation is smaller than the "plasma frequency" ω_p.

in the literature to make an experimental determination of heat transfer modes are based on transfer models. The well-known diffusion model allows the measured total thermal conductivity to be represented as a function of quantities such as radiation temperature, density or porosity of the medium, residual gas pressure and external mechanical load. Extrapolation of thermal conductivity data to a value that is assumed if these quantities are very small sometimes allows model-dependent determination of a single heat flow component.

Why is it that no experimental, i.e. model-independent, method to determine a single heat flow component exists? It is the temperature profile in the medium that couples all heat transport mechanisms (if they depend on temperature).[4] This coupling follows immediately from conservation of energy. Consequently, if a model-dependent determination of a single heat flow component is desired, we have to understand the dependence on temperature of *all* heat transfer modes. If we know this dependence, we can calculate the temperature profile from a solution of an equation that describes conservation of energy, and in a second step calculate radiative heat flow from a solution of the equation of transfer. Frequently this will be an iterative procedure.

Accordingly, this work contains a number of chapters that are normally not included in a radiative transfer volume: we will briefly summarize a minimum of information about temperature dependence of variables and parameters that describe gaseous and solid conduction heat flow and absorption and scattering properties.

If nonradiative conductivity components do not depend on temperature and if the optical thickness of the medium is large, see (1.3-5) for its definition, a very simple determination of a single component is possible: using a mean radiation temperature calculated from the boundary temperatures of the medium, a plot of total thermal conductivity versus this radiation temperature yields a linear relationship (Sect.3.4.5).

---

[4] In a fictitious medium that does not absorb but only scatters radiation, the radiation field is completely decoupled from the temperature profile if its extinction coefficient is temperature-independent. However, this case is only of marginal interest since all real substances absorb at least part of the incident radiation.

If the optical thickness of the medium is large at all wavelengths, the medium is completely nontransparent. As will be outlined in Sect.1.3, a large optical thickness does not indicate that *no* radiation travels through the medium. Examples will be given in Sects.3.4.5 and 4.2, demonstrating that if radiation emitted from boundaries into a dispersed medium has a mean penetration distance of only a few hundred micrometers, radiative heat flow can amount to up to 50% of the residual total heat flow through this medium even if its thickness is in the order of centimeters.

1.2 The Key Role of Radiative Transfer

The key role of radiative transfer manifests itself in at least three ways if the optical thickness of the medium is large:

- All approximate solutions (closed form or numerical) to the equation of transfer converge to the diffusion model solution, which appropriately describes radiative transfer in nontransparent dispersed media, i.e. quantities describing *nonlocal* (radiative) interactions reduce to *local* differential expressions.
- Application of the diffusion model to results of calorimetric measurements allows a separation of radiation from other heat flow components.
- If we apply the diffusion model expression to the separated radiative component, a mean extinction coefficient is obtained that can be compared with spectral extinction coefficients *and* predictions of the Mie theory (if they are wavelength averaged). Thus the extinction properties of a medium can be tackled from three independent positions.

Finally, residual nonradiative (e.g., gaseous or solid conductive) heat flow components extracted from calorimetric measurements can be compared with predictions of appropriate models.

The unique place occupied by nontransparent media in the whole variety of substances is obvious: the diffusion model solution of the radiative transfer problem is, so to speak, the kernel that must be common to all serious attempts of solving the equation of transfer (a few of them will be reviewed in Chap.3). It is encouraging that they indeed converge to the diffusion model solution if the optical thickness of the medium becomes very large. A numerical study that includes temperature dependent thermal conductivity or temperature dependent extinction coefficients or anisotropic scattering is extremely complex if the medium is partially transparent. With the diffusion model applied to nontransparent media, these calculations are easy to perform, and furthermore the results are ready for comparison with calorimetric experiments.

The important step that has made this kind of analysis possible involves a transformation of the integro-differential equations of conservation of energy and of radiative transfer into simple, locally defined differential expressions, i.e. radiative transfer too is considered a conduction process. Spectroscopic and numerical methods are the tools used to decide whether this approach is valid. The corresponding simple criterion is that spectral extinction coefficients have to be as large as possible at all wavelengths.

For these reasons, this work has been confined to radiative transfer in nontransparent media. The most important argument for this restriction was clearly the need for comparison of theoretical data with calorimetric experiments. In addition, nontransparent media are not the exception but rather the rule in the whole catalogue of dispersed materials. A thorough discussion of radiative transfer in partially transparent materials, e.g. exotic substances such as aerogels, requires a separate presentation.

1.3 Definition of Extinction Coefficient, Optical Thickness and Nontransparency

Almost all models that we will use here for the determination of radiative flow in dispersed media originate from astrophysics. Although this is at first sight rather surprising, the solution of the classical radiative transfer problem, i.e. the formulation of the law of darkening of the Sun, is the solution of a *continuum* problem. Application of the same, well established mathematical methods to radiative transfer problems in dispersed media is certainly justified, provided these media can also be considered a *continuum* with

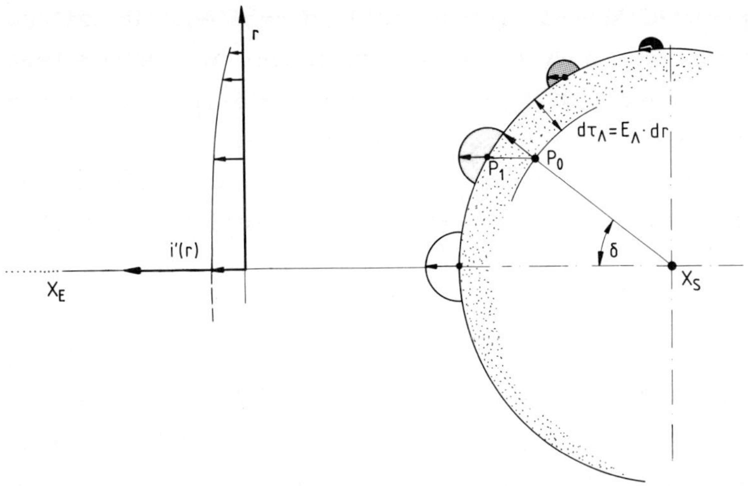

Fig.1.1. Law of darkening of the Sun. The length of the vectors $i'(0,\delta)$ (i.e. the radii of the small circles, or their shading) indicates intensity of black body radiation emitted diffusely at the position P_1 on the Sun's surface. If δ increases, the distance $P_0P_1 = dr/\cos\delta$ and thus the optical thickness $d\tau$ seen by a beam emitted at P_0 increases. x_S and x_E denote coordinates of centers of Sun and Earth

respect to radiation-particle interactions. From a historical point of view, investigation of the law of darkening has led most effectively to an understanding of radiative transfer processes and of the formation of radiative equilibrium.

Let $i_\Lambda'(\tau_\Lambda = 0,\delta)$ denote the directional intensity of a beam that emerges at an angle δ (Fig.1.1) from a point P_0 located near the surface of the Sun [Unsöld 1968 p.111][5]

$$i_\Lambda'(0,\delta) = \int_0^\infty I_\Lambda'(\tau_\Lambda^*)\cdot\exp(-\tau_\Lambda^*/\mu)\cdot d\tau_\Lambda^*/\mu \,. \tag{1.1}$$

In (1.1), Λ denotes the wavelength, and δ is the angle between a line connecting the centers of the Sun (x_S) and the Earth (x_E) and the line x_SP_0. We use $\mu = \cos\delta$. The source function $I_\Lambda'(\tau_\Lambda)$ describes the intensity of radiation emitted and scattered along the optical path

[5] All primed intensities are directional quantities.

at an optical depth (thickness) τ_Λ.[6] In local thermal equilibrium (no scattering), $I_\Lambda'(\tau_\Lambda)$ is a well-known function of temperature (Planck's law). Measurement of $i_\Lambda'(\tau_\Lambda=0,\delta)$ then yields information on the temperature profile in the upper layers of the Sun's atmosphere.[7]

Let us return to the general case. Equation (1.1) is a solution of the "equation of radiative transfer" [Siegel and Howell 1972 p.689]

$$\frac{di_\Lambda'(\tau_\Lambda)}{d\tau_\Lambda} = - i_\Lambda'(\tau_\Lambda) + I_\Lambda'(\tau_\Lambda) \tag{1.2}$$

in a plane medium of infinite thickness. It is assumed that the source function contains only continuous, i.e. thermal, radiation. Note that at this stage this is the only assumption made about the medium's internal structure, i.e. nothing is specified with respect to a possible ratio of absorption and scattering cross sections or the presence of other, nonradiative heat flow mechanisms. If the latter were present, the source function would have to respond to them by virtue of the temperature profile, which in turn is dependent on all heat transfer modes. In this sense Equation (1.2) is universal: it describes radiative transfer through a gas in a high temperature kiln as well as through a glass melt or through a cryogenic thermal insulation, provided these media are continua with respect to radiation-particle interactions.

A medium can be considered a continuum if the geometrical dimensions of its constituents are small compared with the wavelength of the incident and absorbed/emitted and scattered radiation (if this condition is not fulfilled, the laws of classical geometrical optics

---

[6] The more $I_\Lambda'(\tau_\Lambda)$ increases with increasing τ_Λ, the more radiation originating at great optical depths penetrates to the Sun's surface. Darkening is thus increased because i_Λ' assumes a constant value $i_\Lambda'(\tau_\Lambda=0,\delta=\pi/2) = I_\Lambda'(0)$ at the Sun's limb. Although i_Λ' is already a direct measure of the strength of the source function, the $x=\cos\delta$ approximation of Eddington and Barbier [Unsöld 1968 p.114] relates $i_\Lambda'(0,\delta)$ more closely to I_Λ' by $i_\Lambda'(0,\delta) = I_\Lambda'(\tau_\Lambda=\cos\delta)$.

[7] This is true only if the wavelength Λ in (1.1) is different from the position of a spectral line. Observation of spectral lines in a star's atmosphere signalizes the absence of thermodynamic equilibrium. Because absorption at these spectral positions is very strong, emitted intensities stem from upper (i.e., cooler) layers of a star's atmosphere.

and "view factors" describe radiative *exchange* between large absorbing/emitting and reflecting bodies).

If a medium is a continuum, it is possible to define an "optical thickness" that is responsible for an exponential decay of intensity $i_\Lambda'(0)$ incident on a sample at the position s=0

$$i_\Lambda'(s) = i_\Lambda'(0) \cdot \exp\left[-\int_0^s E_\Lambda(s^*)\cdot ds^*\right] . \tag{1.3}$$

Equation (1.3) is the well-known Lambert-Beer law. In differential formulation

$$di_\Lambda'(s) = - i_\Lambda'(s)\cdot E_\Lambda(s)\cdot ds , \tag{1.4}$$

this law is immediately recognized as a special case of the equation of transfer (1.2). The two equations become identical if the source function I_Λ' vanishes.

In (1.3,4) the quantity $E_\Lambda(s)$ denotes a spectral extinction coefficient which, in principle, can depend on temperature. In general, E_Λ is composed of absorption and scattering contributions, see (1.17).

The integral

$$\tau_\Lambda(s) = \int_0^s E_\Lambda(s^*)\cdot ds^* \tag{1.5}$$

gives explicitly the definition of spectral optical thickness, τ_Λ, at the geometrical position s, with reference to $s^* = 0$. If D denotes the total (finite) geometrical thickness of a medium, its optical thickness $\tau_\Lambda(D)$ will be denoted in the following as $\tau_{0,\Lambda}$ or, if the medium is "grey", simply as τ_0.[8] With respect to continuum properties, the definition of τ_Λ reflects an aspect of the structure of $E_\Lambda(s)$: it is sufficient that the integral in (1.5) exists, i.e. $E_\Lambda(s)$ need not be a continuous function. In other words, the number (per

[8] In a "grey" medium, optical parameters such as E or τ (not $m = n-i\cdot k$!) do not depend on wavelength.

unit volume) of constituents in a medium is allowed to vary locally. Even particle geometry may fluctuate.[9]

We are now in a position to define quantitatively nontransparency. In analogy to the definition given previously in Footnote 1, nontransparency will be defined with respect to the attenuation of a beam of original intensity $i_\Lambda'(0)$. Attenuation of a beam is described by (1.3). A beam is the more attenuated, the larger the magnitude of the integral $\tau_{0,\Lambda} = \int E_\Lambda(s) \cdot ds$, see (1.5), in the exponent of (1.3). Attenuation of a beam thus depends only on optical thickness $\tau_{0,\Lambda}$. We will speak of a nontransparent medium if it has an optical thickness of at least 15. According to (1.3), a medium with an optical thickness $\tau_{0,\Lambda}>15$ attenuates an original beam of intensity $i_\Lambda'(0)$ to a value $i_\Lambda'(\tau_{0,\Lambda})/i_\Lambda'(0)$ below 10^{-6} (the reason why $\tau_{0,\Lambda}>15$ has been chosen for definition of nontransparency will be become finally obvious in Sect.3.1, Fig.3.2).

Equation (1.4) is the usual definition of the extinction coefficient E_Λ. A measurement of the attenuation of a beam by a very thin layer using a detector that is mounted exactly in the beam axis thus yields E_Λ directly. Note that no information is available from such a measurement on the relative contributions of absorption and scattering to the (total) extinction coefficient E_Λ. An attenuation measurement that is based on Beer's law always yields an exponential decay of $i_\Lambda'(0)$ regardless of whether the medium absorbs or scatters radiation.

The attenuation measurement performed in this way is correct provided no absorbed/emitted or scattered radiation falls on the detector. However, since all real substances at a temperature above absolute zero absorb and emit and scatter radiation, this condition is in

[9] Chan and Tien [1974a] regards a medium as homogeneous if particle diameters are small compared with sample thickness, and homogeneity is considered essential for establishing a continuum. As a consequence, the reader might be tempted to consider a medium homogeneous also if it consisted of particles of different chemical composition (with particle dimensions small compared with wavelength). However, absorption and scattering cross sections and thus extinction coefficients depend not only on particle geometry (strictly speaking: on the ratio of a characteristic particle dimension and wavelength) but also on complex refractive index, i.e. on composition. Note that (1.3-5) do not require homogeneity to be fulfilled.

many cases only approximately fulfilled. This is especially the case if the substance has strong anisotropic forward scattering properties, which happens if particle diameters are large compared to the wavelength of incoming radiation. Part of the intensity that was taken by the extinction process from the original beam is thus redirected toward the detector. The detector then receives an intensity that is larger than $i_{\Lambda}'(\tau_{0,\Lambda})$ so that application of (1.4) would lead to an underestimate of E_{Λ}. In order to correct for redirected intensity contributions, (1.4) has to be modified by an additional term. Since the redirected contributions seem to be emitted by virtual radiation sources in the interior of the medium, they are accounted for by introduction of the source function $I_{\Lambda}'(\tau_{\Lambda})$ in (1.2). If $I_{\Lambda}'(\tau_{\Lambda})$ were known, an attenuation measurement would thus deliver again the correct E_{Λ}.

The situation is more complicated if we regard transmission measurements. These experiments are related to the penetration of radiative flow. Radiative flow incorporates not only the residual radiation but also absorbed/emitted and scattered radiation that after reaching the opposite side of a layer emerges into the hemisphere. Calculation and interpretation of transmission coefficients (Sect. 3.1) is thus far more difficult compared to the simpler Beer's law. This law considers only one direction for incoming and residual beam. Now we need a mathematical relation between parallel or other modes of incoming radiative flow, optical thickness and hemispherically transmitted flow. Since an integration over the hemisphere has to be carried out, transmission measurements are, in addition to mathematical interpretation, difficult to perform.

Although the original intensity decays completely in a nontransparent medium, absorbed/emitted and scattered flow establish a nonvanishing radiative flow to heat sinks within the medium or to cold boundaries. These contributions become very small if the medium shows strong absorption but they decay considerably more slowly if scattering is predominant (Sect.3.1). Thus a medium can be translucent even if it is nontransparent (a medium is translucent if the sources of incoming radiation cannot be identified by the observation of transmitted radiation; Fig.3.2 shows an interesting example).

No radiation reaches an opposite boundary if the medium is nontransparent and if the source function vanishes.

It is well known that the scattering cross section of a spherical particle is proportional to $1/\Lambda^4$. As a consequence, the extinction coefficient could become small in the far infrared if the medium had only weak absorption properties in this range of wavelengths. Strong absorption at large wavelengths is thus a necessary condition for nontransparency.

1.4 Definition of a Dispersed Medium

In the following, we will adopt usual definitions of a "dispersed" medium. If a (solid, liquid or gaseous) substance is so finely and insolubly distributed in a homogeneous medium that it is in a considerably higher state of energy, by virtue of a large surface area, than its compact phase, the system will be considered a dispersed medium. Dispersed media are usually classified into colloid dispersed and coarsely dispersed systems according to the number of atoms that build up a single "particle". In aerosols, foams, emulsions, suspensions, solid- and hydrosols, particles are composed of 10^3–10^9 atoms. These media belong to colloid dispersed systems. Coarsely dispersed media comprise besides fog, smoke, foams, emulsions and suspensions also clouds, natural and artificial dusts, lattices, pastes, paints, sand, concrete, stones, and others. The single constituents of these systems consist of more than 10^9 atoms. Most colloid and coarsely dispersed media are polydispersed; (a *mono*dispersed medium is usually regarded as being composed of constituents that have identical shape and dimensions. However, from the radiative point of view, the medium is monodispersed if in addition the optical properties of its constituents are identical, that is if they have the same surface structure, density, chemical composition and orientation). A dispersed medium is not necessarily porous.

In cold and dense plasmas, stationary thermodynamic equilibrium is, as in neutral gases, quickly reached by a variety of energy exchanging collision processes, thus leading to an unambiguous definition of (plasma) temperature. Particle velocity and temperature are correlated by the Maxwell distribution. A deviation from this classical concept will appear in Sect.2.2.5: low gas density and the accommodation effect call for a modification of gas temperature. Hot and low-density plasmas allow the formation of only *partial* thermodynamic equilibrium, i.e. with respect to particular groups of particles (atoms, electrons, ions). For the total plasma, a temperature in the conventional sense can no longer be defined. Only neutral particle, electron and ion temperatures exist, and these may deviate strongly from each other. It is also impossible to use an unambiguously defined source function in the transfer equation (1.2) since different optical thicknesses would apply to different radiation sources (spectral lines, recombination continuum, and bremsstrahlung; the latter is the overwhelming contribution in a hot plasma). If the density decreases further, the concept of partial thermodynamic equilibrium and of neutral particle, electron and ion temperature fails completely if the lifetime of the plasma is too short. For these reasons, the methods described below are not applicable to radiative transfer in hot and low-density plasmas (nor to interstellar radiation fields). However, it is not because of the lack of thermodynamic equilibrium, since we will investigate radiative transfer also in strongly scattering dispersed media where incoming radiation is never completely thermalized.

Our discussion will thus be confined to those radiation transfer problems that involve emitted or scattered continuous, spontaneous radiation in the source function.

For the description of radiative transfer, it is sufficient to assume a dispersed (regular or irregular) distribution of geometrical locations in three-dimensional space where nothing but the boundary conditions for the solution of Maxwell's equations have to be observed, i.e. changes of refractive index at these locations. However, the work reviewed below regularly considers *solid particles* that are surrounded by vacuum or by a gas. Nevertheless, (1.2-5) together with calculations of E_Λ by Mie theory have a much wider applicability. For example, the reverse problem (scattering of radiation by holes dispersed in a solid matrix, i.e. scattering by bubbles instead of droplets) can be accounted for simply by considering the reverse of the refractive index [van de Hulst 1981 pp.220-221].

We will further restrict the review to aggregates of spherical and cylindrical solid particles. Generally, it will be assumed that these particles are randomly distributed in space. For cylindrical particles, the maximum freedom allowed in the orientation will be parallel to each other and arranged in planes. The particles may be amorphous or crystalline, and may have either finite or zero electrical conductivity. They may either be completely separated from each other or have direct physical contacts with neighbours (point or surface contacts) or may be linked by chemical bonds.

With few exceptions, we will assume that the particles are essentially dry, i.e. particle surfaces and voids are only partially covered or filled with adsorbed gases or with liquids, respectively. Admission of a liquid that completely surrounds isolated particles introduces no fundamental difficulty: from the pure radiative view, the liquid is simply considered by definition of a *relative* refractive index. Alteration of heat transfer by conduction can be estimated by means of cell models (Sect.2.1.1). Convection is frequently accounted for by introduction of an "effective" thermal conductivity. The "indirect" consequences of these alterations for radiative transfer again follow from conservation of energy. However, the problem of predicting contact heat flow between neighbouring particles has not been satisfactorily solved. In principle, fluctuations in contact heat flow that are caused by adsorbed gas layers or interstitial liquids, for example, are correlated in the same manner as before with alterations in radiative heat transfer. Little experimental work has been published, making it difficult to ascertain whether well-established theoretical models for contact heat flow can yield realistic estimates if contact zones are altered.

Although we are primarily involved with solid particles, we will not assume that their coordinates in space are fixed. The results discussed will remain valid if a constant average particle clearance and a constant total number of particles per unit volume are conserved and if a cloud of particles that is allowed to change its shape exhibits a large optical thickness in all directions and at all times.

Concerning the temperature profile in the medium, we will assume that the profile is defined exclusively by heat transfer characteristics, boundary conditions and thermal capacity.

The occurrence of additional heat sources or sinks in the medium, e.g. chemical reactions, nuclear decay, phase transitions that either consume or deliver additional energy, will not be included in this discussion. Again this is not a severe limitation because in nontransparent media we can always find a coordinate system that transforms a heat source or sink into an appropriate boundary condition.

For simplicity, it will further be assumed that the shape of the particles remains constant, apart from possible local deformations in the contact zones. Particles or particle clusters are expected to retain their individuality and are subject to elastic or inelastic deformations only in the case of high mechanical load or high temperatures. Melting and sintering will not be included in this review.

Materials that are used for continuous microporous thermal insulations are fully covered by this sequence of definitions. We will review experimental work performed with these materials that has culminated in developing new and improving existing quantitative methods for determination or prediction of heat transfer modes. These methods are applicable to a whole variety of dispersed materials if they conform to the above definitions.

1.5 Definition of Thermal Conductivity

Before we proceed with a discussion of heat transfer in dispersed media, we will briefly reflect on the definition of thermal conductivity in an arbitrary medium. Like other factors of proportionality, the thermal conductivity λ is usually considered a quantity that must be independent of the experimental setup used for its measurement. Following Fourier's empirical law, in a stationary state we have

$$\dot{q} = \dot{Q}/A = \frac{\lambda \cdot (T_1 - T_2)}{D} \tag{1.6}$$

for the case of a planar medium of surface A and thickness D. In this relation, $\dot{Q}$ denotes the total heat flow through the medium, caused by a uniform temperature gradient $-dT(x)/dx = (T_1-T_2)/D$. As a consequence, λ denotes a "true" total thermal conductivity, which is a constant.

As a rule, $\dot{Q}$ and λ are composed of several partial heat flow or partial thermal conductivity components $\dot{Q}_i$ or λ_i, respectively. In a thermal insulator, the heat flow components $\dot{Q}_i$ comprise gaseous heat flow $\dot{Q}_{Gas}$, solid conduction heat flow $\dot{Q}_S$, and radiative heat flow $\dot{Q}_{Rad}$. The component $\dot{Q}_{Gas}$ can be composed of conductive and convective contributions. The total heat flow $\dot{Q}$ is always the sum of its components:

$$\dot{Q} = \dot{Q}_{Gas} + \dot{Q}_S + \dot{Q}_{Rad} \;. \tag{1.7}$$

The same applies to λ and its components: With the given gradient $(T_1-T_2)/D$, we have from (1.7)

$$\lambda = \lambda_{Gas} + \lambda_S + \lambda_{Rad} \;. \tag{1.8}$$

However, it can be a rather crude approximation to calculate $\dot{Q}_{Gas}$, $\dot{Q}_S$ and $\dot{Q}_{Rad}$ or λ_{Gas}, λ_S and λ_{Rad} independently of each other because this can lead to a violation of energy conservation (Sect.3.4.2). Furthermore, before we are allowed to write $\lambda = \lambda_{Gas}+\lambda_S+\lambda_{Rad}$, we have to check if all components λ_i really exist. We will consider a heat flow process a *conductive* process if a thermal excitation (e.g., a vibrational state) travels through a medium stepwise and if the steps are very small in comparison with sample thickness or another characteristic dimension. A heat conduction process is thus a *local* event, because it depends on a local temperature gradient. For a temperature gradient to exist, it must be possible to divide a medium into very small volume elements with the condition that they are not empty. If they are not empty, each volume element is filled with a corresponding mass element. Each mass element can be considered a virtual temperature sensor at the position x_i that indicates a local temperature $T(x_i)$. Obviously the gradient $dT(x)/dx$, i.e. an infinitesimal quantity, is the better represented by $\Delta T(x)/\Delta x = [T(x_i) - T(x_{i+1})]/(x_i-x_{i+1})$, the closer x_i and x_{i+1} are spaced. An upper limit for Δx is given by a mean penetration distance that is found from statistical arguments for a particular excitation to propagate. If we consider an emission process, photons that have a mean penetration distance ℓ_m of, say, some hundred micrometers, will in reality suffer many interactions with particles within this distance (otherwise ℓ_m could not be defined as a statistical quantity). As a consequence, Δx must be smaller than ℓ_m. There is a limitation also for ℓ_m. For radiative transfer to be treated as a conduction process, ℓ_m must be much smaller than sample thickness, according to the definition of conductivity given above.

If $dT(x)/dx$ exists, it is possible to rewrite (1.6) as

$$\dot{q} = - \lambda[T(x)] \cdot \frac{dT(x)}{dx} , \tag{1.9}$$

which defines a local thermal conductivity $\lambda[T(x)]$ that may be a constant for all x, or depend on T(x). As will be seen later, the final case is the realistic one.

We now come to an important and far-reaching conclusion: total thermal conductivity exists only if *all* its components λ_i exist. It is not possible to define a total thermal conductivity for a container that is filled with a transparent gas. Rather, a gaseous conductivity may exist if the pressure is high enough that the mean collision distance between gas molecules is small compared with the wall separation. Therefore, "thermal conductivity" of transparent gases is at the most a *partial* conductivity, and values of λ_{Gas} given in the literature must have been cleaned from nonlocal, i.e. radiative, contributions (compare [Tsederberg 1965 p.88] on the work of Wilner and Borelius). The walls of the container "see" each other if the optical thickness of the gas is small. As a consequence, a large optical thickness at all wavelengths is imperative for the existence of total thermal conductivity.[10]

[10] Caps and Fricke [1983] consistently regards the quantity λ not as a 'true' conductivity when it was calculated using Fourier's law from data obtained with a partially transparent SiO_2 aerogel. Likewise the quantities λ that Fine et al. [1980] measure with partially transparent insulators are not conductivities in the proper sense. The same applies to Pelanne [1979,1981], to Bentsen et al. [1984], and to Braun et al. [1983], although the latter find a conductivity for transparent liquids that does not depend on the thickness of the sample.

If this condition is not fulfilled, λ can depend on the experimental setup (for instance thickness of sample or thermal emissivity of walls). This can easily be seen as follows.

If a completely transparent medium is placed between planar parallel walls, $\dot{Q}_{Rad}$ per unit surface area A is given by the well-known law [Siegel and Howell 1972 p.279]

$$\dot{q}_{Rad} = \frac{\dot{Q}_{Rad}}{A} = \frac{\sigma \cdot (T_1^4 - T_2^4)}{1/\epsilon_1 + 1/\epsilon_2 - 1} . \tag{1.10}$$

Note that $\dot{Q}_{Rad}$ does not depend on the wall separation D. In this equation, σ is the Stefan-Boltzmann constant, T_1 and T_2 denote the (constant) temperatures of walls, and the $\epsilon_{1,2}$ are corresponding thermal emissivities (it is assumed that the walls emit radiation according to Planck's law and that the $\epsilon_{1,2}$ are independent of temperature). One is not allowed to apply the definition of λ_{Rad} given in (1.8), i.e.

$$\lambda_{Rad} = \dot{q}_{Rad} \cdot \frac{D}{T_1 - T_2} \tag{1.11}$$

to (1.10). If this were to be done, then because q_{Rad} given in (1.10) is independent of D, we would have

$$\lambda_{Rad} = \dot{q}_{Rad} \cdot \frac{D}{\Delta T} = \frac{\sigma}{1/\epsilon_1 + 1/\epsilon_2 - 1} \cdot \frac{T_1^4 - T_2^4}{T_1 - T_2} \cdot D = \mathrm{const} \cdot D ,$$

i.e., λ_{Rad} would depend on the thickness D.

In the case of a nonvanishing optical thickness τ_0, $\dot{q}_{Rad}$ assumes approximately [Sparrow and Cess 1966 p.231]

$$\dot{q}_{Rad} = \frac{\sigma \cdot (T_1^4 - T_2^4)}{1/\epsilon_1 + 1/\epsilon_2 - 1 + 3\tau_0/4} . \tag{1.12}$$

Remembering that $\tau_0 = E \cdot D$, we would have again

$$\lambda_{Rad} = \frac{\sigma}{1/\epsilon_1 + 1/\epsilon_2 - 1 + 3E \cdot D/4} \cdot \frac{T_1^4 - T_2^4}{T_1 - T_2} \cdot D = \frac{C_1}{C_2 + 3E \cdot D/4} \cdot D$$

where E is the extinction coefficient and $C_{1,2}$ are constants. The ratio $C_1/(C_2+3E \cdot D/4)$ can be expanded in a series. This yields

$$\lambda_{Rad} = \text{const} \cdot (1 - \frac{C_3}{D} + \frac{C_3^2}{D^2} - \frac{C_3^3}{D^3} + \ldots)$$

$$= \text{const} \cdot \left[1 - \frac{C_4}{\tau_0} + (\frac{C_4}{\tau_0})^2 - (\frac{C_4}{\tau_0})^3 + \ldots\right]$$

where the C_3 and C_4 are given by $C_3 = (4/3)C_2/E$, $C_4 = C_3 \cdot \tau_0/D$. Obviously, the dependence of λ_{Rad} on thickness decreases with increasing E and D because τ_0 (i.e., nontransparency) increases. If E and D decrease, transparency increases, and λ_{Rad} is definitely dependent on thickness. To make a provisional distinction between conductivities of nontransparent and transparent media, in the following, we will speak of a (total or partial) conductivity that depends on thickness (or on thermal emissivity $\epsilon_{1,2}$, see below) as a "pseudo-conductivity" λ'. Whereas the effect of thickness is strong in λ_{Rad}', its importance is reduced in the total λ' if the gaseous or solid heat flow components are described by "true" conductivities λ_{Gas} and λ_S.

Another conductivity that is mentioned frequently in the literature is "apparent conductivity". This expression is used if a strong curvature of T(x) or if large temperature gradients dT(x)/dx are expected in a planar medium near its boundaries. Calculating conductivity as usual from measured total $\dot{Q}$ gives only approximate values since local values of dT(x)/dx can differ strongly from $-(T_1-T_2)/D$ (Fig.1.2). The existence of large temperature gradients near the boundaries is due to weak thermal coupling of the medium to its surroundings, e.g. to the walls (in the very special case of a nonconducting medium, even a temperature jump will be observed, see

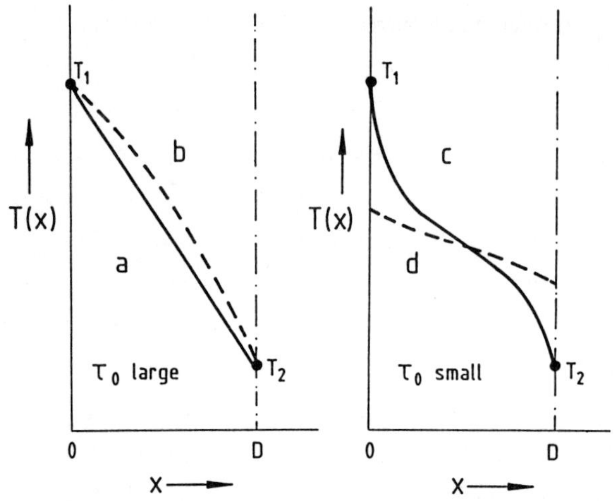

Fig.1.2. Stationary temperature profiles T(x) (schematic) versus thickness in nontransparent (curves a and b) and transparent (curves c and d) planar media. Case a: "*true*" total thermal conductivity λ exists and is a constant. Case b: temperature dependence of conductivity component λ_{Rad} is responsible for a (small) curvature of T(x), i.e. the medium has only an *apparent* total thermal conductivity. Case c: small optical thickness τ_0 and conduction properties of the medium cause large temperature gradients near the boundaries. Because of the small τ_0, the total (apparent) conductivity is thickness dependent and is thus a *pseudo*-conductivity. In cases a-c, dT(x)/dx exists for all x. Case d: small optical thickness and vanishing conduction properties of the medium cause temperature jumps at the boundaries, dT(x)/dx does not exist for all x, and the medium's total conductivity is a *pseudo*-conductivity

Sect.3.4.2). Weak thermal coupling to the wall can happen if the optical thickness of the medium is small. Therefore the real background for "pseudo-conductivity" and "apparent conductivity" seems to be the same. However, if one speaks of apparent conductivity, it is implicitly assumed that dT(x)/dx, though strongly varying, exists everywhere in the medium. In a medium for which a pseudo-conductivity has been calculated, dT(x)/dx might not exist at all, e.g. in a Dewar vessel. On the other hand, "true" conductivity is associated with dT(x)/dx that exists everywhere in the medium and equals or is at least very close to $-(T_1-T_2)/D$.

It is thus clear from (1.9) that all local values $\lambda[T(x)]$ equal $(Q/A)\cdot D/(T_1-T_2)$ = const in the case of "true" conductivity. But $\lambda[T(x)]$ varies according to the variation of dT(x)/dx if the medium only has a pseudo (or apparent) thickness-averaged conductivity. In a Dewar vessel, a local $\lambda[T(x)]$ does not exist at all.

It is left as an open problem to what extent frequently used pulsed photoacoustic techniques for the measurement of absorption coefficients in highly transparent materials are subject to uncertainties resulting from thickness-dependent pseudo-conductivities.

A possible dependence of λ_{Rad} on $\epsilon_{1,2}$ can be demonstrated in the same way. Let $C_3 = (4/3)C_2/E = (4/3)(1/\epsilon_1+1/\epsilon_2-1)/E$ be large. This happens if E is small. If D is also small, λ_{Rad} depends on $C_2 = 1/\epsilon_1+1/\epsilon_2-1$. Observations of this effect have been made by Fine et al. [1980] and Kunc et al. [1984]; the latter found a reduction of the pseudo-conductivity of borosilicate glass by about 20% if wall emissivity is altered from 0.95 to 0.05. That emissivity dependent pseudo-conductivity really exists already follows from Viskanta [1965].

The same observations will be made in the case of an evacuated multifoil insulation, which can certainly be considered nontransparent (otherwise it would be impossible to reduce radiative losses with this type of insulation to the smallest values that have ever been achieved; these "superinsulators", however, do not belong to dispersed media). Evacuation and the use of spacers between metallic foils makes $\dot{Q}_{Gas}$ and $\dot{Q}_S$ negligibly small, i.e. $\dot{Q}_{Rad}$, although very small too, is the dominant residual heat transfer mode. If $\dot{Q}_{Rad,0}$ denotes radiative flow without foils and N is the number of foils, we have

$$\dot{q}_{Rad,N} = \frac{\dot{Q}_{Rad,N}}{A} = \frac{\dot{Q}_{Rad,0}}{A} \cdot \frac{1}{N+1}$$

if wall and foil emissivities are equal [Kaganer 1969a p.34]. Again, $\dot{q}_{Rad}$ is independent of wall separation D, i.e. if λ_{Rad} is calculated as before, it would depend on thickness D of the stack of foils. This example shows that a simple division of media into transparent and nontransparent species does not yield a definite answer as to whether thermal conductivity depends on thickness or not.

A pseudo-conductivity dependent on thickness (or wall emissivity) is thus not necessarily a consequence of transparency but rather due to a dominating radiative component in the total heat flow. Conversely, a conductivity that is independent of these quantities does not necessarily involve nontransparency but indicates prevailing "true" conduction mechanisms.

1.6 Conservation of Energy Defines Temperature Profile

If heat sources and sinks are absent, conservation of energy requires

$$\nabla \cdot \dot{q} = 0 \tag{1.13}$$

where $\dot{q}$ is the total heat flow density

$$\dot{q} = \dot{q}_{Gas} + \dot{q}_S + \dot{q}_{Rad} \; . \tag{1.14}$$

Equation (1.13) couples the components $\dot{q}_i$ via the temperature profile in the medium. Let us assume that the nonradiative conductivity components λ_i can be represented by a "true" conductivity λ_{Cond}. If the medium is planar, homogeneous, absorbing and emitting, isotropically scattering, grey and conductive, (1.13) reads in more detail [Viskanta 1965]

$$N_1 \cdot \frac{d^2\Theta}{d\tau^2} = (1 - \Omega) \cdot [\Theta^4(\tau) - \eta(\tau)/4] \; . \tag{1.15}$$

In this equation, N_1 is the usual conduction/radiation parameter [Siegel and Howell 1972 p.631]

$$N_1 = \lambda_{Cond} \cdot E/(4\sigma \cdot n^2 \cdot T_1^3) \tag{1.16}$$

and Θ denotes the dimensionless stationary temperature, $\Theta = T/T_1$ (T_1 is the "hot" wall temperature). The quantity n^2 is the effective value of the squared real part of the complex index of refraction, see the discussion in Sect.3.4.6. The quantities Ω and $\eta(\tau)$ denote albedo of single scattering and total radiative flow that is incident on a unit volume, respectively.

If the extinction coefficient E is composed of an absorption coefficient A and a scattering coefficient S, we have

$$E = A + S \; . \tag{1.17}$$

In nondispersed, highly pure solids, E usually equals A. The scattering contributon S to E arises only if fluctuations of the refractive index come into being, e.g. if the solid is dispersed.

The albedo Ω is defined as

$$\Omega = S/E \tag{1.18}$$

so that $S = \Omega \cdot E$ and $A = (1-\Omega) \cdot E$. Equation (1.15) thus correlates the difference between absorbed incident $A \cdot \eta(\tau)/4$ and emitted radiation $A \cdot \Theta^4(\tau)$ with the energy that flows to or from the volume element by nonradiative processes [left side of (1.15)].

Equation (1.15) delivers the temperature profile $\Theta(\tau)$, which can be influenced by a variation of calorimetric *and* optical quantities such as λ_{Cond} or Ω. Since emitted radiation and conductive heat flow depend on $\Theta^4(\tau)$ or $d\Theta/d\tau$, respectively, it is immediately obvious that a variation of the *calorimetric* quantity λ_{Cond} influences not only conductive but also *radiative* heat flow, and that a variation of the *optical* quantity Ω influences not only radiative but also *conductive* heat flow. More explicitly we have, for instance at $\tau=0$ and assuming wall emissivities $\epsilon_{1,2} = 1$,

$$\frac{\dot{q}_{Rad}(0)}{\sigma T_1^4} =$$

$$1 - 2 \cdot \left\{ \Theta_2^4 \cdot E_3(\tau_0) + \int_0^{\tau_0} [(1-\Omega) \cdot \Theta^4(\tau^*) + \frac{\Omega}{4} \cdot \eta(\tau^*)] \cdot E_2(\tau^*) \cdot d\tau^* \right\} \tag{1.19}$$

$$\frac{\dot{q}_{Cond}(0)}{\sigma T_1^4} = - N_1 \cdot \frac{d\Theta}{d\tau}(0) \; . \tag{1.20}$$

In (1.19), the $E_n(\tau)$ denote the well-known exponential integral functions whose use greatly facilitates solid angle integration (isotropic scattering assumed) when calculating $\dot{q}_{Rad}$ from directional intensities. For a description of the properties of $E_n(\tau)$ compare [Chandrasekhar 1960 pp.373-374]. The $E_n(\tau)$ go rapidly to zero if $\tau \to \infty$. Therefore, terms in (1.19), which are multiplied by $E_n(\tau)$ can be interpreted as being only locally active. The extent to which (1.19) is coupled to a locally defined conductivity law is thus given by the magnitude of the optical thickness τ_0.

The influence of all important parameters (not only Ω and λ_{Cond}) on the temperature profile and heat flow components in a grey medium was studied by Viskanta [1965] and earlier (Sect.3.3.2).[11] However, anisotropic scattering has been considered in the literature only recently, and with very few exceptions existing work is confined to the application of temperature independent parameters. The discussion in Chaps.2 and 3 will demonstrate that a consideration of anisotropic scattering and temperature dependent calorimetric and optical parameters is imperative for an understanding of heat flow components.

According to (1.15) the coupling of heat flow components in the temperature profile cancels exactly if the medium does not absorb but only scatters radiation (Ω=1) (and if N_1 is a constant). In this case the temperature profile is exactly linear because (1.15) reduces to $d^2\Theta/d\tau^2 = 0$. If N_1 is large, i.e. if λ_{Cond} *and* E are large, the temperature profile is only approximately linear (the influence that N_1 exerts on temperature profiles is clearly seen from the experimental data given in Fig.3.13, Sect. 3.3.5). If on the other hand, N_1=0 and Ω<1, then (1.15) reduces to $\Theta^4(\tau) = \eta(\tau)/4$, which describes radiative equilibrium. Since the factor (1-Ω) in (1.15) cancels under these conditions, the temperature profile no longer depends on albedo. As a consequence, absorption and isotropic scattering can be considered as

[11] The definition of conduction/radiation parameter N_1 is not unique in Viskanta's work (sometimes calculated curves are labeled with $1/N_1$). We will follow the definition of N_1 given by Siegel, but the square of the refractive index will be included as is done by Viskanta. Except for comparisons with experimental results, it will suffice to assume $n^2 = 1$ in the derivations reviewed.

equivalent extinction processes if the medium has no conduction properties. If $N_1 > 0$, this equivalence is broken (Fig.3.8e, Sect.3.3.2). In the case of an arbitrary value of Ω ($0 \leq \Omega \leq 1$) and for large values of τ_0 the solutions of (1.19) *converge* to an expression for $\dot{q}_{Rad}$ that follows from the diffusion model of radiative transfer (Sect.3.4).

The "additive approximation" assumes that the $\dot{q}_i$ can be calculated independently from each other. Within this approximation, we have for large τ_0

$$\dot{q} = \dot{q}_{Cond} + \dot{q}_{Rad} = \lambda_{Cond} \cdot \frac{T_1 - T_2}{D} + \frac{4\sigma \cdot n^2 \cdot (T_1^4 - T_2^4)}{3\tau_0} \tag{1.21}$$

and

$$\lambda = \lambda_{Cond} + \lambda_{Rad} = \lambda_{Cond} + \frac{4\sigma \cdot n^2}{3E} \cdot \frac{T_1^4 - T_2^4}{T_1 - T_2} \, . \tag{1.22}$$

Clearly, λ given in (1.22) is a *thickness-averaged* quantity. As a consequence for calculating λ from (1.22) it has to be assumed that the temperature profile is to a good approximation linear, and that there is perfect or at least very strong thermal coupling of the medium to the wall.

We will come back to the applicability of the additive approximation and to the diffusion model in Sect.3.4.2. With respect to the exact and approximate solutions of (1.15,19) it should be emphasized that pure isotropic scattering is not a regular but an exceptional case.

If a *local* description of heat flow components $\dot{q}(x)$ and conductivities $\lambda_i(x)$ is desired, it is necessary to know the temperature dependence of calorimetric and optical parameters. A considerable part of Chapter 2 is devoted to a review of temperature dependence of parameters that are necessary to solve (1.15,19).

1.7 Three Independent Methods to Determine Extinction Coefficients

Experimental investigations that are reviewed in this volume comprise (calorimetric) measurements of thermal conductivities and optical experiments to detect extinction, absorption and scattering properties. With regard to thermal conductivity measurements, those experiments that are performed under well-defined mechanical external load will be considered in more detail. Devices that use cylindrical or spherical geometry (compare [McElroy and Moore 1969] for a survey; [Rohatschek 1976]) do not allow a variation of mechanical boundary conditions. Since evacuated thermal insulators and geological structures are usually subject to mechanical load, it is necessary that an apparatus for measurement of thermal conductivity can simulate this condition. In addition, it has been suggested that peg-supported vacuum insulations should be developed [Nowobilski 1979; Ziegenbein and Reiss 1980] that can be prepared only in planar geometry. A large majority of all existing devices for measurement of thermal conductivity seem to be of plane-parallel geometry. However, only very few of them allow measurements under definite mechanical boundary conditions that are exactly maintained during the experiment.

The accuracy of thermal conductivity measurements greatly depends on uniformity of temperature level in the heating and reference plates (one or more guard rings are usually applied to simulate an infinitely extended isothermal plane). For example, an inhomogeneity of only 1 K can introduce an experimental error of 3-4 per cent in λ [Büttner et al. 1983]. Thus the need for accuracy in temperature measurements cannot be overemphasized. Measurement of heat flow through insulators that integrate metal foils is of particular difficulty because of a high lateral thermal conductivity [Grunert et al. 1969b].[12]

Besides determination of extinction coefficients and albedo, optical experiments involve measurements of scattering phase functions and investigations of how independent scattering[13] may be affected by particle clearance.

In the case of a large optical thickness, (1.22) shows that the *calorimetric* quantity "thermal conductivity" is coupled to the optical quantitiy "extinction coefficient".

Finally, application of the Mie theory yields a third independent method to determine extinction coefficients. If spectral refractive indices can be chosen from experiment, it is possible to achieve agreement within 10% between the results of these three approaches. This is a rather surprising result because application of the Mie theory is dependent on ideal particle geometry and independent scattering. The first assumption is hardly fulfilled in real substances. Independent scattering, however, seems to dominate at least if porosity exceeds 0.95 and if the particles are (approximately) spherical.

1.8 Radiative Contribution to Heat Flow at Low Temperature

Heat transfer in nonevacuated, dispersed thermal insulators is dominated by conduction processes; at high gas pressure, convection is increasingly important (conduction and convection usually amount to more than 75% of the total heat flow). If the medium is evacuated, radiative heat flow can be the most important transfer mechanism

[12] In an evacuated multifoil insulator consisting of 40-50 Al foils, the ratio of thermal "pseudo"-conductivity perpendicular and parallel to temperature gradient is about 10^5–10^6. Because of different cross sections that multifoil *insulation* (parallel to temperature gradient) and foils (perpendicular to temperature gradient) assume, the corresponding ratio of heat fluxes in these directions is smaller: if 50 Al foils, each with a cross section of 10 μm by 1 m are assumed, with 30 mm thickness of the multifoil stack and a conductivity of 200 W/(m·K) of pure Al and a conductivity of 0.5 mW/(m·K) of the evacuated multifoils, the ratio is still about 200. For this reason, a considerable reduction of Al thickness is necessary. One way of achieving this is through evaporation of Al (0.05 μm thickness) on Mylar.

[13] If we consider N identical particles per unit volume, we speak of independent scattering if the total extinction coefficient E is N times the extinction cross section of one particle; this finding may depend on wavelength (Sect.2.4.3).

even if the medium is at cryogenic temperatures. Kaganer [1969a pp.77-78] measured thermal conductivity of aerogels and perlite: although the wall temperatures T_1 and T_2 were only at 290 K $\leq T_1 \leq$ 300 K and 77 K $\leq T_2 \leq$ 90 K, radiative contribution accounted for more than 80 per cent of the total thermal conductivity (this result is achieved by extrapolating the measured conductivity to T = 0; we will come back to a thorough description of this method in Sect.3.4). Because of the high infrared transparency of aerogels below 8 μm wavelength, a situation may arise where this calorimetric method is not applicable (i.e., at high temperatures; see [Scheuerpflug et al. 1985]).

2. Quantities Needed to Formulate the Equations of Energy Conservation and Radiative Transfer

In the previous chapter it has been seen that a solution of the equation of transfer in real substances regularly depends on the existence or nonexistence of other modes of energy flow through a medium and their possible temperature dependence. As a consequence, we should consider transport models that describe conductive (and in principle, convective) heat flow *before* we proceed to a review of solutions of the radiative transfer equation.

2.1 Parameters and Functions, Cell and Continuum Models

In the following, quantities such as wavelength, extinction coefficient, gaseous thermal conductivity and others that enter theoretical heat-transport models will be formally grouped into variables and parameters or parametric functions. The reader will soon agree that it is helpful for organization of the following review to clearly separate between methods to evaluate physical dependencies and mathematical methods to solve the equation of transfer.

To give an example, with respect to an attempt at solving the equation of transfer the extinction coefficient E_Λ is not an appropriate variable for this purpose. It is a very complicated function of physical parameters (particle geometry, particle density and clearance, refractive index; the latter depends on chemical structure of the substance, on wavelength, on polarization of incoming radiation and on temperature). If we know from whatever source numerical values of E_Λ, then the optical thickness τ_Λ is defined. From this instant, τ_Λ instead of E_Λ can be regarded as merely a *mathematical* variable, and it remains to be investigated how the equation of transfer can be solved as a function of τ_Λ (this problem is indeed purely mathematical).

Sections 2.2 and 3 are devoted to a review of the temperature dependence of the parametric functions $\dot{Q}_{Gas}$ and $\dot{Q}_S$, which are needed for a solution of the energy conservation equation [(1.15) or its equivalents]. Section 2.2 involves in some depth a study of traditional literature. Section 2.4 deals with parametric functions that are necessary for the determination of $\dot{Q}_{Rad}$. This includes a survey on Mie theory and dependent scattering.

Solutions of the equation of transfer are reviewed in Chapter 3.

2.1.1 Cell and Continuum Models

Before we start with a discussion of the temperature dependence of $\dot{Q}_{Gas}$ and $\dot{Q}_S$, we will briefly review the basic properties of cell and continuum models.

It has been mentioned in Chap.1 that radiative heat flow will be treated exclusively by application of continuum models (for a survey on history and manifold of continuum models and their counterpart, cell models, and their importance for radiative transfer see [Vortmeyer 1979]). Scanning electron microscopy of aggregates of powders or fibres (Sect.4.1) gives no indication that application of classical[1] cell models would probably be successful: arrangement e.g. of particles in SiO_2 aerogel (Fig.2.11) or fluctuations of the particle geometry and coordination number in a bulk, are in clear contradiction to idealized assumptions such as those of Deissler and Eian [1952], Woodside [1958] (identical spherical particles in a cubic array), or Dietz [1979] (hexagonal array).

---

[1] Classical cell models usually assume a highly symmetrical arrangement of constituents in three-dimensional space. However, with regard to radiative flow, this is only part of the reason that they fail in the case of a microporous medium: in reality, it is the assumption of opacity of the constituents that has to be corrected if the particle dimensions are of the order of the wavelength of incoming radiation. It is justified to consider fractal models [Mandelbrot 1983] as cell models. However, the nonradiative heat propagation obeys diffusion laws that are different from the usual diffusion model [Boccara and Fournier 1987].

Although Woodside [1958] successfully describes thermal conductivity of snow (considering a diffusive flow of water vapour through the bulk of snow particles) and Deissler and Eian [1952] reproduced measured heat flow through a bulk of spherical MgO, SiO_2, SiC or steel and lead particles, by calculation, agreement found in these papers is certainly ambiguous because radiative flow is not considered. This is not a peculiarity of Deissler's and Woodside's work. According to Vortmeyer [1979] neglect of radiative flow that penetrates through the void spaces in a particle array is *typical* for classical cell models. Although this assumption anticipates large optical thickness (because the solid constituents are regularly considered opaque), classical cell models are, at least for pure geometrical reasons, less suited for nontransparent media.

Attempts were made in the literature to improve cell models by the inclusion of radiation that flows through void spaces or by the calculation of radiative exchange between opposite surfaces, respectively. Since particle dimensions in beds are usually large compared with wavelength, radiative flow is thus calculated in terms of "macroscopic" reflection and thermal emission coefficients. This macroscopic limit of the scattering of a plane wave is reflected by radiative Nusselt numbers [Wakao and Kato 1969]:

$$Nu_{Rad} = \frac{4\sigma \cdot T^3 \cdot d}{(2/\epsilon - 1) \cdot \lambda_S} = \frac{\alpha_{Rad} \cdot d}{\lambda_S} . \qquad (2.1)$$

In this equation, d denotes particle diameter, and ϵ is the thermal emission coefficient of the particle surface. We can regard α_{Rad} as a heat transfer coefficient that is the radiative analogue to the heat transfer coefficients for free convection.[2]

Equation (2.1) explicitly states that radiative heat flow is considered in terms of *surface* properties, i.e. it is assumed that radiation does not penetrate into the particle's interior. In reality, small particles are not at all opaque. On the contrary, scattering

---

[2] The coefficient α_{Rad} can be calculated according to $\alpha_{Rad} = [2B+\epsilon \cdot (1-B)] \cdot 4\sigma \cdot T^3/[(2-\epsilon) \cdot (1-B)]$, where B denotes a geometrical constant [Wakao and Vortmeyer 1971; Vortmeyer 1979].

properties of a small particle can be calculated correctly only if electromagnetic fields inside the particle are accounted for (consideration of wave functions inside a small dielectric particle is especially important). It is not the main obstacle for an application of cell models to properly describe radiative transfer in finely grained substances that real substances are far away from the idealized assumptions on particle geometry, space symmetry, and purity (these assumptions could eventually be fulfilled from a statistical viewpoint). Rather, it is a mismatch of wavelength of incident radiation to particle dimension that cannot be healed by statistics.

An intermediate position between the pure cell and pure continuum models is, besides [Chan and Tien 1974a], occupied by Laubitz [1959]: a pure cell model that was originally designed by Russell [1935] for calculating thermal conductivity is for purely empirical reasons extended by the inclusion of a diffusion-like radiative term (we will come back to Laubitz's work in Sect.2.2.6 and to the diffusion model in Sect.3.4).

A concept which is inherent in continuum models is that of a mean free path between sequential interactions. This concept has proved to be useful in a variety of fields (energy loss of charged particles due to ionization, scattering of low energy neutrons in the moderator of a nuclear reactor, collisions between gas molecules at medium and high gas pressure, phonon scattering by imperfections, light scattering in nontransparent media). An important benefit is associated with the mean free path concept in thermal physics: if a mean free path ℓ_m exists for a particular heat transfer mode and if ℓ_m is small compared to a characteristic dimension, this mode can be described as a conduction process. Considering radiative transfer, the corresponding mean free path $\ell_{Rad,\Lambda}$ is an unambiguous function of the spectral extinction coefficient $E_\Lambda(s)$ at a position s in the medium. We have [Siegel and Howell 1972 p.414]

$$\ell_{Rad,\Lambda}(s) = \int_0^\infty E_\Lambda(s)\cdot\exp\left[-\int_0^s E_\Lambda(s^*)\cdot ds^*\right]\cdot s\cdot ds \ .$$

If E_Λ does not depend on s, this reduces to

$$\ell_{Rad,\Lambda} = \frac{1}{E_\Lambda} \ . \tag{2.2}$$

In this case, the optical thickness $\tau_{0,\Lambda}$, (1.5), is equal to the number of mean free paths $\ell_{Rad,\Lambda}$ in a particular medium. It is obvious that the individuality of a particle or a "particle-particle phalanx" characteristic for cell models is completely lost in the statistical expression for $\ell_{Rad,\Lambda}$ given by (2.2). On the other hand, we will see in Sect.2.3 that the application of cell models is widespread in the case of contact heat flow.

2.2 Heat Flow Through a Gas in a Two-Phase System

2.2.1 Basic Relations for Calculating Gaseous Conductivity

Thermal conductivity λ_{Gas} of a gas (phase a) that flows between the solid particles (phase b) of a dispersed medium can be calculated from kinetic gas theory. In an ideal gas of infinite extension in space, we have (see standard textbooks on kinetic theory of gases, e.g. [Loeb 1961 or Kennard 1938 p.164])

$$\lambda_{Gas} = (1/3)C_V \cdot \rho_{Gas} \cdot \ell_{Gas} \cdot v = \lambda_0 \ . \tag{2.3}$$

In (2.3), C_V denotes specific heat capacity at constant volume, ρ_{Gas} denotes gas density, ℓ_{Gas} is the mean free path between two collisions of gas molecules, and v is the mean velocity of the molecules. The mean free path ℓ_{Gas} can be estimated according to Sutherland's well-known formula [Sutherland 1893]

$$\ell_{Gas} = \frac{C_1}{1 + C_2/T} \ ; \tag{2.4}$$

this expression accounts for attractive interactions between colliding molecules. The mean velocity v is given by a Maxwellian distribution [Kennard 1938 pp.45-50] and depends on $T^{1/2}$. The gas temperature T is unambiguously defined if ℓ_{Gas} is small compared to the mean distance between the walls that enclose the gas volume, which was assumed in (2.3) to be infinitely large.

If a gas volume is enclosed in a pore, i.e. in a *finite* volume, (2.3) holds as long as ℓ_{Gas} is small compared to pore diameter δ. As a consequence, we have to discuss methods that yield a reliable estimate for the pore diameter. Furthermore, it should be ascertained to what extent rotational and oscillatory degrees of freedom of the gas molecules as a function of temperature, alter the specific heat C_V in (2.3) [Jakob 1964 pp.73-76]. The dependence of C_V on these states is usually taken into account with a correction factor ϵ so that $\epsilon \cdot C_V$ becomes the correct representation of the specific heat. In addition, we must be aware of the magnitude and temperature dependence of the accommodation coefficient α [Knudsen 1911; Kennard 1938, Chaps. 180-183].

Before we discuss the parameters δ, ϵ, and α we note the expression for λ_{Gas} defined in a single pore of a dispersed medium. A "temperature jump" [Kennard 1938 pp.311-315] occurs at the pore wall because in the immediate vicinity of the wall ℓ_{Gas} is no longer small compared to the corresponding distance (Fig.2.1). To account for the

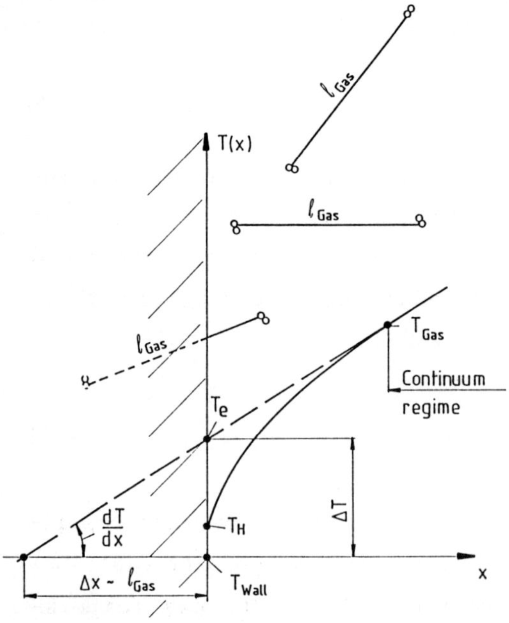

Fig.2.1. Temperature jump ΔT between T_{Gas} of the continuum regime and T_{Wall}. In a diatomic gas, the jump distance Δx is approximately equal to ℓ_{Gas} [Wutz 1965 p.29]. The temperatures T_e (linearly extrapolated temperature from T_{Gas}) and T_H (accommodated temperature) are needed in Fig.2.3

temperature jump, which effectively extends the wall distance, (2.3) has to be altered. In the shortest notation [Kaganer 1969a p.6] we have

$$\lambda_{Gas} = \frac{\lambda_0}{1 + 2\beta \cdot Kn} \; . \tag{2.5}$$

Here β denotes a weight factor for the Knudsen number

$$Kn = \frac{\ell_{Gas}}{\delta} \; . \tag{2.6}$$

The factor β comprises the above mentioned corrections for specific heat and accommodation

$$\beta = \frac{2\epsilon}{\kappa+1} \cdot \frac{2-\alpha}{\alpha} \; . \tag{2.7}$$

In this equation, κ is the usual ratio $\kappa = C_p/C_V$ of specific heat at constant pressure C_p and at constant volume C_V.

Figure 2.2a,b [Kaganer 1969a] illustrates how the well-known de-

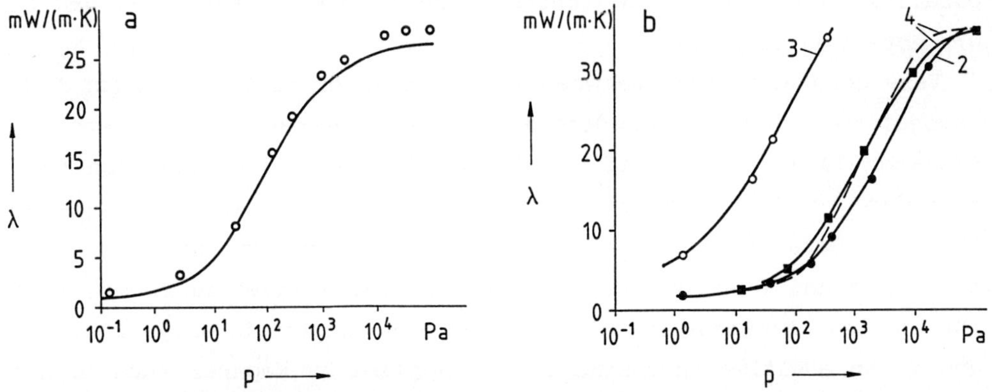

Fig.2.2. (a) Variation of thermal conductivity λ of spherical particles (expanded perlite sand) with residual gas pressure p (air) at low temperature. Open circles denote experimental data. The full curve was calculated using a two-phase expression for total λ [Kaganer 1969a p.18]. (b) Variation of thermal conductivity λ of fibrous particles with residual gas pressure (air). Data (symbols) are taken at low mean temperatures. Fibre diameters are 0.7 μm (curve 2), 3.6 μm (curve 4) and 15.2 μm (curve 3). Solid and dashed curves are calculated from (2.8) using only the second term (Fig.2.12) [Kaganer 1969a p.76]

pendence of $\dot{Q}_{Gas}$ and λ_{Gas} on the Knudsen number or residual gas pressure is reflected in fine spherical or cylindrical particle media.

Note that (2.5) only *modifies* (2.3) by a factor $1/(1+2\beta\cdot Kn)$, i.e. the concept of heat conduction characterized by molecular collisions is not altered except for a small region of space near the pore wall; in particular, (2.5) does not introduce free molecular conduction. However, according to (2.5) if gas pressure is constant λ_{Gas} now depends on pore diameter δ, i.e. λ_{Gas} is a pseudo-conductivity if β and Kn are large.

We will first discuss estimates of the pore diameter. If it is assumed that micropores of particles (index 1) and macropores between particles (index 2) can be considered separately in an expression for λ_{Gas} that is adapted from (2.5), we obtain [Kaganer 1969a p.72]

$$\lambda_{Gas} = \frac{\lambda_1}{1 + 2\beta\cdot \ell_{Gas}/\delta_1} + \frac{\lambda_2}{1 + 2\beta\cdot \ell_{Gas}/\delta_2} \;. \tag{2.8}$$

Equation (2.8) assumes that gaseous heat flow through pore 1 occurs parallel to the heat flow through pore 2. The numerators λ_1 and λ_2 are empirical constants.

Very good agreement between (2.8) and experimental data for SiO_2 aerogel and other fine powders at cryogenic temperatures is reported [Kaganer 1969a p.72ff.] if a solid conduction and a radiation term are added to (2.8). Unfortunately, this equation cannot be proved in this way because SiO_2 aerogel is highly transparent at wavelength between 2 and 7 µm. If the thickness of the samples is 20 mm, then the optical thickness in this range of wavelengths is only between 1 and 2. An additive approximation as applied by Kaganer thus is not admissible. Furthermore, the pore diameters δ_1 and δ_2 entering (2.8) are not precisely defined. Kaganer [1969a p.14] and Deimling [1984 p.87] applied a "hydraulic" diameter of the void spaces for an estimate of δ_2. If Π denotes the porosity, they find $\delta_2 = (2/3)\cdot[\Pi/(1-\Pi)]\cdot d$, d being the diameter of a grain. Although hydraulic diameters are very useful in fluid dynamics to account for complicated cross sections in ducts, their application in this case is doubtful because macropores do not necessarily have closed surfaces. In addition, for a porosity close to 1, we would find δ_2 near infinity, which is not realistic.

Verschoor and Greebler [1952] calculate δ_2 from collision probability between gas molecules and solid particles: $\delta_2 = C \cdot d/(1-\Pi)$, with C a constant and d the diameter of spherical or cylindrical particles. Like the relation given by Kaganer, this formula is valid only for medium values of porosity Π. If we consider not only fluctuations in particle geometry and particle diameter, but also mixtures, e.g. coarse-grained, radiation scattering opacifiers (with diameters between 2 and 5 μm) and very fine particles such as SiO_2 aerogel (about 0.005 to 0.01 μm in diameter), and agglomerations, the resulting uncertainties in both formulae with respect to pore diameter are obvious. However, Verschoor's formula seems to be more suited to handling these situations.

We will not consider here the experimental methods to determine pore diameters or pore diameter distributions, e.g. by capillary condensation methods and application of Kelvin's equation ([Defay et al. 1966 pp.217ff.] and references cited therein). Where data for λ_{Gas} are given as a function of Knudsen number, it should be checked how the pore diameters were determined.

The equipartition theorem allows a first estimate of how translatory, rotational, and vibrational excitations of the gas molecules contribute to the specific heat. Each translatory or rotational degree of freedom should possess an energy of $k \cdot T/2$, each vibrational degree $2k \cdot T/2$; the factor 2 in the latter case results from the independent kinetic and potential energy contributions to vibrational motion. However, at low temperatures molecular vibrations are "frozen", so that the equipartition theorem, i.e. classical statistical theory, no longer applies (see below). In any case, the total energy U and, since $C_V = (\partial U/\partial T)_V$, the total specific heat C_V of an ideal gas are given by the sum of the contributions from all excitations: $C_V = C_{V,Trans} + C_{V,Rot} + C_{V,Osc}$. At very high temperatures, another contribution $C_{V,El}$ of electronic excitations must be considered but this will be neglected here. Thus, according to statistical theory we have at low temperatures, i.e. without the inclusion of vibrations, $C_V = 3R/2$ for monatomic and $C_V = 5R/2$ for diatomic ideal gases. Modifications to these simple estimates have to be introduced not only with respect to the onset of vibrations at higher temperatures. Application of the arithmetic mean square velocity has led to the contribution $k \cdot T/2$ per degree of translatory motion. However, molecules that carry higher energy, i.e. have a higher temperature, are faster than those of lower energy load because temperature is proportional to v^2. Thus a correction factor ϵ_{Trans} has to be applied (see below.) If similar weights ϵ_{Rot} and ϵ_{Osc} are introduced, we have a total correction factor ϵ.

The correction factor ϵ to C_V that accounts for the weights of rotational and vibrational states of gas molecules in addition to translatory motion reads explicitly [Jakob 1964 p.74]

$$\epsilon = \frac{\epsilon_{Trans} \cdot C_{V,Trans} + \epsilon_{Rot} \cdot C_{V,Rot} + \epsilon_{Osc} \cdot C_{V,Osc}}{C_{V,Trans} + C_{V,Rot} + C_{V,Osc}} \,. \qquad (2.9)$$

The velocity distribution of monatomic gases predicts $\epsilon_{Trans} = 5/2$ [Chapman 1912]. For the ϵ of He, Ne, Ar, Kr and Xe, Eucken [1940] reported very good experimental agreement with this value.

An average of ϵ_{Osc} taken over the three orientations of the major axes yields $\epsilon_{Osc} = 3/2$ [Eucken 1913]. If rotational and intramolecular oscillatory states of gas molecules could be treated as independent of translatory motion, then $\epsilon_{Rot} = \epsilon_{Osc} = 1$. However, if the axis of a diatomic gas molecule is for instance parallel to the temperature gradient, a central collision with another gas molecule can excite translatory *and* oscillatory motion. These two different states of excitation are thus not necessarily independent of each other. Nevertheless, Eucken [1940] and Jakob [1964 pp.75-76] reported good agreement for the ϵ of gases with negligible oscillatory energy (N_2, O_2 at T = 0°C) and of gases with a nonvanishing oscillatory energy such as CO, NO and Cl_2 when they compared predictions of (2.9) using $\epsilon_{Rot} = \epsilon_{Osc} = 1$ with experimental values. On the other hand, they attribute too small theoretical values for ϵ of H_2 at temperatures $0 \le T \le 400$°C and of CCl_4 and C_2H_6 at T = 0°C to a certain coupling of translatory and oscillatory energy at elevated temperatures. Theoretical values that are too large are found for H_2O vapour at $100^oC \leq T \leq 400^oC$, and for SO_2 and NH_3 at $T = 0^oC$; these are interpreted as the result of an electrical dipole moment.

Classical theory cannot correctly describe the C_V of H_2. Because of its very small moment of inertia, excitation of rotational states is not possible at lower temperatures; it requires a relatively large amount of thermal energy, i.e. at least room temperature.

At low temperatures the quantity $k \cdot T$ per vibrational degree of freedom is small compared with $\hbar \cdot \omega$ or multiples thereof. As a consequence, no vibrational excitations exist. The predictions of classical theory are thus overestimates at low and even at medium temperatures. Only if $k \cdot T >> \hbar\omega$ will the full vibrational spectrum be excited. A temperature dependence of ϵ_{Osc} is thus a direct consequence of the Boltzmann distribution. The quantum theory of specific heat and the equipartition theorem are thus in agreement only at elevated temperatures.

Equation (2.9) then seems to allow satisfactory estimates of the influences of molecular excitations on C_V.[3] According to Eucken the experimentally determined ϵ values of H_2O vapour and air remain nearly constant if the temperature is increased, whereas the ϵ of CO_2 seems to decrease and that of H_2 to increase at $T=0^oC$. Eucken's result for the ϵ of H_2O vapour, however, is not confirmed in later publications: Tsederberg [1965 pp.79-81] reported from experimental investigations of Tarzimanov that the ϵ of H_2O vapour increases with room temperature at all vapour pressures (the same reference also claims temperature dependent ϵ for monatomic gases from investigations of Zaytseva.) From Eucken's and Tsederberg's work, it seems more likely that ϵ really does depend on temperature. Since H_2O vapour is one of the most important gases we will consider ϵ generally to increase with increasing temperature, in accordance with the more recent report of Tsederberg.

The factor $2\epsilon/(\kappa+1)$ in (2.7) results from a derivation of the temperature jump $\ell_{Gas} \cdot dT/dx$ [Kennard 1938 pp.312-315]. Since κ is nearly

---

[3] Eucken [1940] recommended a couple of additional ways to improve the accuracy, also [Jakob 1964 p.76].

temperature independent, any temperature dependence of $2\epsilon/(\kappa+1)$ is caused solely by the factor ϵ.

2.2.2 Temperature Dependence of the Accommodation Coefficient

Next we have to check a possible temperature dependence of the accommodation coefficient α. This coefficient describes the extent to which gas molecules that collide with wall molecules adapt to the wall temperature. According to its definition, the accommodation coefficient should depend strongly on the surface properties of a pore wall. Like the correction factor ϵ described in Sect.2.2.1, the accommodation coefficient is composed of translatory, rotational and vibrational contributions.

Blodgett and Langmuir [1932] report values of α for H_2 between 0.54 and 0.1 to 0.2 for tungsten surfaces when the surface is either pure, or covered with H_2, or oxidized, respectively. This would imply higher values of α for very regular surfaces. However, Roberts [1930] and Michels [1932] found increasing values of α when an annealed wire cooled down gradually. According to these observations, adsorbed gases seem to increase α, contrary to the findings of Blodgett.

Keesom and Schmidt [1936] showed that at low temperatures, the α of He, H_2, Ne and N_2 on glass decreases with increasing temperature. Raines [1939] showed that α of He on a clean, gas-free Ni surface increased with increasing temperature whereas it decreased when the Ni surface was already covered with gas. The α taken at the gas-covered surface were a factor of 5 to 10 larger than those resulting from the gasfree surface. Thus the trends of α with temperature observed by Roberts, Michels and Keesom were confirmed by the work of Raines, and later by Klett and Irey [1968] for air and nitrogen on copper.

A dependence of α on mass was reported by Roberts [1933] and by Keesom and Schmidt (1936). The relation

$$\alpha = \frac{2m_1 \cdot m_2}{(m_1 + m_2)^2} \tag{2.10}$$

derived by Baule [1914], also [Kennard 1938 p.325], yields increasing

α with increasing m_2 if $m_2<m_1$ (in this equation m_1 and m_2 denote the masses of molecules of the outermost layers of the wall and of the gas molecules, respectively). This expectation is supported by experimental observations of Wakao and Vortmeyer [1971]. According to these authors, measured and theoretical values of thermal conductivity have been compared for Zr and U particles (190 μm diameter) or spheres of steel (3.18 mm diameter) embedded in H_2, He, N_2, air, CH_4 and Ar (the theoretical values were calculated using a cell model). For a large range of gas pressures Wakao reports values of α=0.3 for the light gases H_2 and He, and α=1 for the heavier gases. Good agreement is found for temperatures up to 30°C.

Recent experiments again illustrate large fluctuations in α [Thomas 1967; Springer 1971; Saxena and Afshar 1985]. The work of Saxena confirms the findings of Roberts and Michels (see above) because an increase of α with increasing gas pressure is observed. The reader will intuitively follow Kostylev [1964] who assumes that several layers of adsorbed molecules on a disperse material at low temperatures are favourable to equalize thermal states of colliding molecules. Since adsorbed layers disappear at elevated temperatures, accommodation will decrease. Thus α should decrease with increasing temperature, as confirmed by Saxena and Afshar [1985], at least up to $T \leq 1100$ K (Fig.2.3).

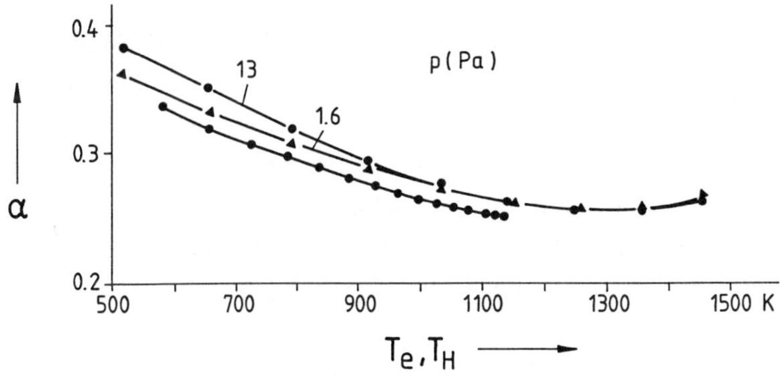

Fig.2.3. Thermal accommodation coefficient α of the argon-tungsten system as a function of temperature and at different gas pressures p. The T_e and T_H refer to Fig.2.1 [Saxena and Afshar 1985]. The two upper curves refer to the low pressure region and the bottom curve to the temperature jump regime

2.2.3 Conclusions for the Temperature Dependence of Gaseous Conductivity

Since ϵ of H_2O vapour increases whereas α generally decreases with increasing temperature, the factor β, (2.7), for H_2O is most likely to increase with temperature. This increase is probably weak. For CO_2, β would be approximately constant. If we introduce the usual relations $\rho_{Gas} = \rho_{Gas}(1/T)$, C_V = const, $v = v(T^{1/2})$ and $\ell_{Gas} = \ell_{Gas}(T)$ into (2.3), we find $\lambda_0 = \lambda_0(T^{1/2})$. More accurate relations for C_V and ℓ_{Gas} use $\epsilon(T)\cdot C_V(T) = \epsilon(T)\cdot C_3(1+C_4\cdot T)$ and $\ell_{Gas} = C_1/(1+C_2/T)$ according to (2.4), where C_1, C_2, C_3, and C_4 are constants. This leads to

$$\lambda_0 = \lambda_0(T^n) \ , \tag{2.11a}$$

$$\lambda_{Gas} = \lambda_{Gas}(T^n) \tag{2.11b}$$

where the exponent n in (2.11a) takes into account the temperature dependencies of C_V, δ, ℓ_{Gas}, v, and ϵ. If ϵ varies linearly with temperature, as is suggested by Zaytseva and Tarzimanov [Tsederberg 1965 pp.80–81], we have at elevated temperatures approximately n = 3/2. An exponent n of approximately 1/2 (as suggested, e.g., by Dushman [1958 pp.46–47]) holds only at lower temperatures because oscillatory and eventually rotational degrees of freedom are frozen, i.e. if $\epsilon\cdot C_V$ = const, and because the temperature dependencies of ρ_{Gas} and ℓ_{Gas} compensate each other in (2.3).

From kinetic gas theory, the exponent n in (2.11b) is again approximately 3/2 if $\epsilon=\epsilon(T)$, if β is a constant and if $2\beta\cdot Kn \gg 1$. However, Tsederberg [1965 p.89] reported experimentally determined values of n for a variety of diatomic, triatomic, and polyatomic gases at elevated temperatures from the work of Zaytseva and Vargaftik. Values of n near 1/2 result only for monatomic and diatomic gases, e.g. $0.7 \le n \le 0.98$ for He, Ne, Ar, Kr, Xe, and Hg vapour or $0.78 \le n \le 0.87$ for H_2, N_2, O_2, and CO. Triatomic gases (H_2O, CO_2, and SO_2) have $1.23 \le n \le 1.48$. Since the agreement with both the simple $T^{1/2}$ or $T^{3/2}$ law is poor, Tsederberg [1965 pp.88–97] analyzes λ_0 at atmospheric pressure in terms of the number of atoms and of the critical temperature T_{Crit} as a reference temperature. If τ denotes T/T_{Crit}, Tsederberg gives the relation

$$\log\left[\frac{\lambda_0(T)}{\lambda_0(T_{Crit})}\right] = C_1\cdot\log\tau + C_2\cdot(\log\tau)^2 + C_3\cdot(\log\tau)^3 \qquad (2.12)$$

which yields good agreement with experiment (C_1, C_2, C_3 are constants). If this relation is used in (2.5) the temperature dependence of λ_0 and λ_{Gas} will be equal only if β is a constant and at elevated temperatures.

Thus we find approximately the same temperature dependence for the (true) thermal conductivity of a gas that is infinitely extended (λ_0) and for the pseudo-conductivity λ_{Gas}' of the same gas enclosed in a pore.

2.2.4 Convection as a Conduction Process

If gas pressure increases, the onset of convection leads to an increase of λ_{Gas} and of total λ. Bett and Saville [1965] report measurements of λ using Al_2O_3 particles (1-3 mm diameter) embedded in Ar. It is shown (Fig.2.4) that the linear relationship $\dot{Q}/(\lambda_{Gas}\cdot\Delta T) \propto Pr\cdot Gr$ holds to within ±10% for natural convection (in this equation Pr and Gr denote Prandtl and Grashoff numbers, respectively; if this relation is extrapolated to high gas pressures, deviations from a linear relationship would denote the onset of turbulence). Figure 2.4

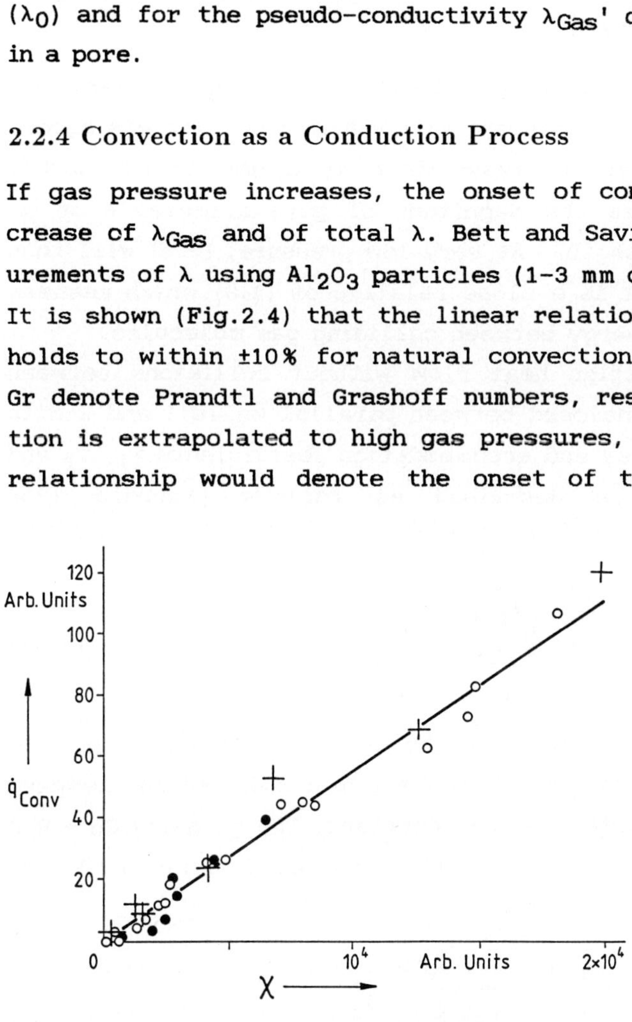

Fig.2.4. Correlation of convective heat loss $\dot{q}_{Conv}$ with the product χ = Pr·Gr given for packed beds of bubbled alumina [Bett and Saville 1965]

illustrates the well-known fact that for laminar gas flow conditions, convective heat transfer can always be interpreted in terms of an effective thermal conductivity λ_{Eff}: if χ=Pr·Gr we have

$$\lambda_{Eff} = (1 + \chi)\cdot\lambda_{Gas} \ . \tag{2.13}$$

A temperature dependence of λ_{Eff} thus can be investigated if in addition the temperature dependence of Pr and Gr is analyzed at elevated gas pressures.

2.2.5 Free Molecular Conduction

We will now discuss the temperature dependence of λ_{Gas} at very low gas pressures. Note that the concept of conductivity that was agreed upon in Sect.1.5 necessarily breaks down if a gas is evacuated, because ℓ_{Gas} soon reaches the magnitude of pore diameter δ or of another characteristic length. At very low pressure, (2.5) will thus become invalid because it is a close relative of (2.3) which assumes a stepwise exchange of energy between colliding gas molecules.

Free molecular conduction heat flow without collisions between the molecules of a gas enclosed between parallel walls 1 and 2 with corresponding temperatures and accommodation coefficients T_1, T_2 and α_1, α_2, respectively, is described as follows [Kennard 1938 pp.315-318]:

$$\dot{q}_{Gas} = \frac{\alpha_1\cdot\alpha_2}{\alpha_1 + \alpha_2 - \alpha_1\cdot\alpha_2}\cdot\rho_{Gas}\cdot\sqrt{\frac{R\cdot T}{2\pi}}\cdot(C_V + R/2)\cdot(T_1 - T_2) \ . \tag{2.14}$$

In this equation $\rho_{Gas} = p/(R\cdot T)$ is the density of the gas (we may use the relation for an ideal gas because T is a modified gas temperature, see below). R denotes the gas constant. The quantity $C_V + R/2$ is equal to $(\kappa+1)\cdot C_V/2$. A formal conductivity can be derived [do not apply (1.6)!] from (2.14):

$$\lambda_{Gas} = \frac{\alpha_1\cdot\alpha_2}{\alpha_1 + \alpha_2 - \alpha_1\cdot\alpha_2}\cdot\frac{\kappa+1}{2}\cdot\frac{C_V\cdot p}{(2\pi\cdot R\cdot T)^{1/2}} \ . \tag{2.15}$$

Like the gas density ρ_{Gas}, the gas pressure p is related to a modified gas temperature T that depends on the accommodation coefficients:

$$\frac{1}{T^{1/2}} = \left(\frac{1}{2}\right)\left(\frac{1}{\hat{T}_1^{1/2}} + \frac{1}{\hat{T}_2^{1/2}}\right), \tag{2.16a}$$

$$\hat{T}_1 = \frac{\alpha_1 \cdot T_1 + \alpha_2 \cdot (1 - \alpha_1) \cdot T_2}{\alpha_1 + \alpha_2 - \alpha_1 \cdot \alpha_2} \quad \text{and} \tag{2.16b}$$

$$\hat{T}_2 = \frac{\alpha_2 \cdot T_2 + \alpha_1 \cdot (1 - \alpha_2) \cdot T_1}{\alpha_1 + \alpha_2 - \alpha_1 \cdot \alpha_2}. \tag{2.16c}$$

It would be favourable if the three quantities $\alpha_1 \cdot \alpha_2/(\alpha_1+\alpha_2-\alpha_1 \cdot \alpha_2)$, κ and C_V could be considered temperature independent because in this case the same temperature dependence of λ_{Gas} results as before: using $p \propto T$ and V = const,

$$\lambda_{Gas} = \lambda_{Gas}(\sqrt{T}) \tag{2.17}$$

with T defined in (2.16a-c). However, since α decreases with T, the factor $\alpha_1 \cdot \alpha_2/(\alpha_1+\alpha_2-\alpha_1 \cdot \alpha_2) = \alpha/(2-\alpha)$ increases with T (using $\alpha_1 = \alpha_2 = \alpha$), and there is also a temperature dependence of C_V. Therefore, the frequently used relations (2.11) with n = 1/2 [Cabannes 1980a,b; Deimling 1984 p.77] and (2.17) are only approximately fulfilled.

Although (2.15) does not describe a stepwise transport of energy between colliding particles, it fulfills the most important consequence that arises from the classical conduction concept: λ_{Gas} as given in this equation does not depend on dimensions of the gas volume [in contrast to (2.5)]. Thus it is not possible to reduce λ_{Gas} by a reduction of pore diameter if the free molecular conduction regime is already achieved. Reduction of λ_{Gas} is possible only by further evacuation.

2.2.6 Two-Phase Thermal Conductivity

We are finally in a position to consider the applicability of frequently used two-phase expressions [Kaganer 1969a pp.7-18] for the thermal conductivity if a gas surrounds a solid phase. Although it is the goal of these models to yield an expression for total thermal conductivity, they usually do not take radiation into account. As a consequence, these models are of limited validity.

The two-phase models are regularly derived from cell models. No conflict arises with the above made restriction to continuum models for radiative transfer: in principle, it is sufficient to check if reported experimental nonradiative heat flow components confirm predictions of theoretical nonradiative two-phase cell models. However, it is an open problem whether experiments can be performed that are completely free of radiative contributions in the total heat flow. As a consequence, the method would be successful only if appropriate models for radiative flow are used simultaneously. Apparently, the first successful attempt to verify a two-phase cell model in this manner was made by Laubitz [1959]. Laubitz used a cell model originally derived by Russell [1935]. Russell considered a matrix composed of gas and solid particles. The gas is treated as a continuous phase. The solid particles are considered obstacles against heat flow (the role of continuum and obstacles can be interchanged). With regard to the prior work of Maxwell and Rayleigh, who for determination of electrical conductivity use idealized highly symmetrical pictures of spherical particles rigorously attached to sites in a cubic lattice, Russell introduced cubic particles in cubic cell volumes. Laubitz further generalized this view by assumption of a statistical distribution of cubic particles in cubic cell volumes and derived a radiation term from this picture. When additionally applying a numerical correction factor to Russell's cell model, Laubitz reports very good agreement between calculation and experiment performed with magnesia and alumina at temperatures between 100 and 1000°C and at normal gas pressure.

The two-phase expressions for thermal conductivity apply λ_{Gas} values that result from (2.5) or (2.15) according to a given residual gas pressure. Note that in addition to the parameters required for these equations, only porosity and conductivity λ_{SM} of the solid material the particles consist of have to be known. These parameters are easily measured. As a consequence, extending the comparison of theory with experiment from pure kinetic gas theory to two-phase models does not increase the uncertainty that is at any rate involved with the entrance parameters for (2.3-16) (the "rigid" models of Deissler and Eian [1952] and Dietz [1979] require only λ_{SM} as additional parameter). The same argument applies to the temperature dependence of the two-phase model parameters.

The problem of how adsorbed gas layers alter two-phase thermal conductivity will be discussed at the end of Sect.2.3.3.

2.3 Contact Heat Flow Through Solid Phase

2.3.1 The Thermal Resistor Concept

Whereas this section deals with contact heat flow through the solid phase of a dispersed medium, a more general treatment of contact conductivity that is not restricted to interacting spherical or cylindrical particles is presented by Fried [1969].

Contrary to the foregoing section we will now discuss methods to calculate *explicitly* pure contact heat flow. Quantitative expressions that have been developed for our purpose regularly apply the thermal resistor model: the total thermal resistance R_S of a dispersed medium is calculated as the sum of a series of locally defined resistance units. Each resistance unit consists of four partial resistors that are switched partly in series, partly parallel.

This atomistic view clearly reflects the principles of cell models (Fig.2.5): one of the partial resistors (1) is defined with respect to the thermal conductivity of the solid material of which the particles are made. If d is a characteristic dimension of a particle parallel to the temperature gradient[4] and A its cross section, then $R_{SM} = d/(A \cdot \lambda_{SM})$.

[4] In the case of a network of fibres, d is the distance between contacts.

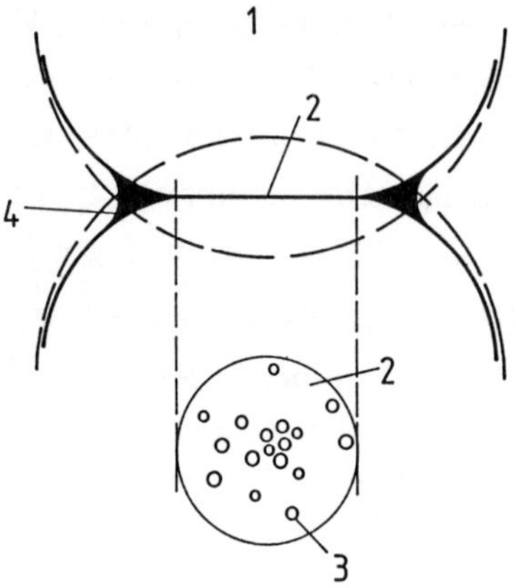

Fig.2.5. Thermal resistance unit (cell) between two spheres. For definitions of partial resistances 1...4 see text. Geometry and dimensions of deformations and contact area are adapted from Landau and Lifschitz [1965]

Another partical resistor (2) takes into account a "macroscopic" contact area between particles. Its size follows from the theory of elasticity, and it defines a contact resistance R_{Cont} that is usually analyzed in terms of a constriction zone (see below). Contribution 3 arises from "microscopic" point contacts located within the macroscopic contact area that are caused by surface roughness. Finally, a film resistance (4) may be important because of a possible contamination. Of the resistances R_{SM} and R_{Cont}, it is the latter that usually dominates. Little information is available on resistances 3 and 4 and their influence on the total R_S.

Whereas R_{SM} is unambiguously defined by d, A and λ_{SM} (see above), there are only approximate expressions for R_{Cont}.

For simplicity, we will consider only circular contact areas. More general contact geometries (elliptical contact zones) are treated in textbooks on theory of elasticity [Landau and Lifschitz 1965].

As mentioned before, the first classical cell models were developed by Maxwell and Rayleigh for a determination of electrical conductivity of a two-phase system. It was only a logical step to restate these results as a solution of the analogous thermal conductivity problem. In a similar manner the thermal contact resistance R_{Cont} has been derived from an already existing expression for its electrical analogue; the electrical resistance of a constriction zone of length b and radius a [Holm 1967 p.16] reads:

$$R^{El}_{Cont} = \left[\rho_{El}/(2\pi\cdot a)\right] \cdot \arctan(b^{1/2}/a) \ . \tag{2.18}$$

In this equation, ρ_{El} denotes the specific electrical resistance of the conductor. If b is very large, we have

$$R^{El}_{Cont} = \rho_{El}/(4a) , \qquad (2.19)$$

Note that (2.18,19) describe the resistance of a single constriction. If two long constrictions are in series, as is the case if they are in contact, the total R^{El}_{Cont} is given by

$$R^{El}_{Cont} = \rho_{El}/(2a) . \qquad (2.20)$$

This result has been experimentally confirmed by Holm to within ±1.5%. The analogous solution to the thermal contact conduction problem (taking into account that $\rho_{El} = 1/\lambda_{El}$) reads

$$R^{Therm}_{Cont} = 1/(2a\cdot\lambda_{Therm}) = 1/(2a\cdot\lambda_{SM}) = R_{Cont} . \qquad (2.21)$$

We have thus found surprisingly simple expressions for partial resistors 1 and 2.

If the ratio d/a of diameter d and contact radius a of a spherical particle equals at least 10 and if two contact patches are diametrically opposed, then $R_{SM}+R_{Cont} = 1.1\ R_{Cont}$ [Kaganer 1969a pp.20-21]. The same reference gives values for d/a of above 20 for thermal insulation materials or above 100 for ordinary granular media. Accordingly, R_{Cont} by far dominates over R_{SM} in these cases.

For resistor 3 (spot contacts because of roughness), the problem could formally be attacked by application of (2.19) to a number N of parallel switched spot resistances. If R^{El}_{Spot} denotes the total spot resistance, we have

$$1/(R^{El}_{Spot}) = \sum_{\nu=1}^{N} 1/(R^{El}_{Spot,\nu}) = \sum_{\nu=1}^{N} \frac{1}{\rho_{El}/(4a_\nu)} \qquad (2.22)$$

if the $R^{El}_{Spot,\nu}$ are independent of each other. However, it is neither clear that we are allowed to take the limit $b \to \infty$ of (2.18) in the case of a minute roughness, nor is it realistic to assume that the independence criterion is fulfilled if the spots are located very closely. Holm [1967 p.22], assuming all radii a_ν equal, gave instead of (2.22)

$$R^{El}_{Spot} = \frac{\rho_{El}}{2\pi \cdot N \cdot a} \cdot \arctan\left[\frac{\sqrt{\ell^2 - a^2}}{a}\right] = R^{El}_{Spot}(N,a,\ell) \ . \qquad (2.23)$$

In this equation, $(\ell^2-a^2)^{1/2}$ is the height of the "a spots", that is, a parameter analogous to the parameter b in (2.18). Since $(\ell^2-a^2)^{1/2}$ can hardly be measured with sufficient accuracy, Holm [1967 p.23] compared two limiting cases: a regular distribution of N single spots that are well separated from each other, and a compact contact surface that is the union of the N spots. If $\hat{d}$ denotes the distance between spots, Holm reported a ratio of $R^{El}_{Spot}(N,a,\ell)/R^{El}_{Cont}$, (2.23,19), of about 0.1 for N=100 and for large values of $\hat{d}/a$. Since a large $\hat{d}/a$ is certainly a realistic assumption, calculation of R^{El}_{Cont} and thus of R^{Therm}_{Cont} from (2.20,21) thus yields an upper limit for the contact resistance. This finding confirms the rather casual assumption that Cunnington and Tien [1972] made on the relative importance of resistors 2 and 3.

However, Goldsmid and Kaila [1980] report for the relation $a \propto p^x$ (p denoting mechanical load) an exponent x of only 0.12 instead of the well-known exponent 1/3 that follows from the theory of elasticity assuming a *closed* contact area (see below). This is not in contradiction to the previous result that stressed the dominance of R^{El}_{Cont} over R^{El}_{Spot}: The small exponent possibly indicates that first the number N of spot contacts increases with increasing p before the radius a starts to increase with $p^{1/3}$ when N reaches a saturation value. On the other hand, Kaganer [1969a p.19] reports a derivation of thermal contact conductivity made by Dulnev and Sigalova who found $\lambda_{Cont} \propto p^{11/18}$, taking into account only "a spots" (note that $\lambda_{Cont} \propto a$). If N = 1 is assumed (that is, if this formula is used to describe the dependence of λ_{Cont} not on R_{Spot} but on R_{Cont}), Kaganer [1969a p.22] reports λ_{Cont} for sand and a glass bead that overestimate experimental values of λ by a factor of about 1.5 to 4. The same occurs if λ_{Cont} is based exclusively on R_{Spot} (i.e., N>1) because in Dulnev's formula $\lambda_{Cont} \propto N^{1/2}$. A pure "a-spot formula" thus yields overestimates of λ_{Cont} whereas Holm's pure R_{Cont} formula (2.19) overestimates R_{Cont}, that is underestimates λ_{Cont}. Thus the contribution of microcontacts to heat flow resistance because of surface roughness (partial resistor c) must not be neglected completely, contrary to the work of Kaganer.

2.3.2 Contact Conductivity of Spheres and Fibres

The above expressions for resistors 1, 2 and 3 are concerned with partial contributions to a resistance unit. Before we discuss the influence of a surface contamination (resistor 4) we will use the expressions for resistors 1 and 2 to review formulae developed for the total thermal resistance of a bulk of spheres or a fibre network.

For a bulk of *spherical* particles, Kaganer [1969a p.21] reported

$$\lambda_{Cont} = 3.7a \cdot \frac{(1 - \Pi)^2 \cdot \lambda_{SM}}{r} . \tag{2.24}$$

We will use Hertz' formula for the contact radius a in the case of elastic deformations [Hertz 1882]

$$a = \left[\frac{3P \cdot r}{4} \cdot \frac{1 - \mu^2}{Y} \right]^{1/3} . \tag{2.25a}$$

In these equations, Π denotes the porosity, r is the radius of the particles, μ the Poisson ratio, P is the load on a single contact, and Y denotes Young's modulus of elasticity. The factor 3.7 in (2.24) arises from the average number N of contacts that a particle experiences with its neighbours. Different approaches for N have been published. Kaganer [1969a p.11] approximated experimental values of N, measured by Kiselev, by the formula

$$N = 11.6(1 - \Pi) . \tag{2.26}$$

If Π is large, for instance $\Pi \simeq 0.9$ as is usual with fine powders, N will probably be underestimated because this expression yields $N \leq 1$ already for $\Pi \geq 0.914$. Meissner et al. [1964] reported the relation

$$N = 2 \exp[2.4(1 - \Pi)] . \tag{2.27}$$

This relation excellently reproduces the experimental values of N of spherical particles in the range $0.25 \leq \Pi \leq 1$. Between $0.25 \leq \Pi \leq 0.45$, another expression [Deresiewicz 1958 p.244] that is based on measurements of Smith et al.

$$N = 26.49 - 10.73/(1 - \Pi) \tag{2.28}$$

yields good agreement with (2.27). A correction of the numerical factor in (2.24) by application of Meissner's results seems appropriate.

Using (2.26), Kaganer [1969a p.19] gives $P = (4\pi \cdot r^2/11.6) \cdot p/(1-\Pi)^2$, which leads to the contact radius

$$a = 0.93\, r \cdot \left[\frac{1 - \mu^2}{Y \cdot (1 - \Pi)^2} \cdot p \right]^{1/3} \tag{2.25b}$$

so that λ can be calculated from (2.24). Note that the "a-spot" are neglected.

For an idealized rectangular lattice of *cylindrical fibres* Kaganer [1969a pp.22-24] derived the following expression

$$\lambda_{Cont} = \frac{16(1 - \Pi)^2}{\pi^2} \cdot \lambda_{SM} \cdot \left[\frac{1}{1.86 \cdot A \cdot p^{1/3}} + \frac{1}{4(1 - \Pi)} \right]^{-1} . \tag{2.29}$$

In this equation, the factor A is used to account for Hertz' contact radius for two crossed cylinders, i.e.,

$$A = \left[\frac{1 - \mu^2}{Y \cdot (1 - \Pi)^2} \right]^{1/3} . \tag{2.30}$$

The specific load is indicated by p. Partial resistors 1 and 2 are given by

$$R_{SM} = \frac{\ell}{\pi \cdot r^2 \cdot \lambda_{SM}} \quad \text{and} \tag{2.31a}$$

$$R_{Cont} = \frac{1}{2a \cdot \lambda_{SM}} \tag{2.31b}$$

where ℓ denotes the distance of contacts, and r is the fibre radius. The R_{SM} and R_{Cont} are switched in series. The "a-spot" are again neglected.

The contact radius for two crossed cylinders is calculated from Hertz's formula again assuming elastic deformations

$$a = \left[\frac{3P \cdot r}{2} \cdot \frac{1 - \mu^2}{Y} \right]^{1/3} . \tag{2.32a}$$

P is again the load on a single contact, $P = (\pi^2/16) \cdot p \cdot d^2/(1-\Pi)^2$ [Kaganer 1969a p.24] so that

$$a = 1.55\ r \cdot \left[\frac{1 - \mu^2}{Y \cdot (1 - \Pi)^2} \cdot p \right]^{1/3} . \tag{2.32b}$$

Kaganer [1969a pp.20-21] showed that for spherical particles $R_{Cont} >> R_{SM}$. With regard to fibres, we have the same relation if the distance ℓ between contacts and the contact radius are small. In both cases it is correct to assume $R_S = R_{Cont}$. Small contact radii are favoured by large Young moduli, that is, hard materials with a highly symmetrical lattice. Although the λ_{SM} of quartz is higher than the corresponding values of substances with a lower degree of lattice symmetry (such as graphite), or amorphous structure (such as glass), the smaller contact radii of quartz particles compensate for the larger λ_{SM}. Therefore, large λ_{SM} will not always lead to large λ_S.

From $R_S = R_{Cont}$ we conclude that $\lambda_S = \lambda_{Cont}$. However, it will be shown in Sect.4.2 that experimental results for λ_S require splitting:

$$\lambda_S = \lambda_{S,0} + \lambda_{SC} \; . \tag{2.33}$$

In this equation, $\lambda_{S,0}$ denotes the part of solid thermal conductivity that is present also in the absence of an external mechanical load, whereas λ_{SC} coincides with (2.24 or 29). It is likely that besides λ_{SM} the parameters Π, μ, Y will also appear in $\lambda_{S,0}$ because of a small residual load that is caused by the weight of the bulk or by thermal stresses.

The temperature dependence of λ_S is thus identified with the temperature dependence of those parameters that enter (2.24 or 29) or even (2.23). Taking into account the "a-spot" would not alter the temperature dependence of λ_{Cont} from both (2.24 and 29) and thus of λ_S.

We will see shortly that the corresponding complete sets of parameters are also identical.

Neglecting the term $1/[4(1-\Pi)]$ in (2.29) we have

$$\lambda_{Cont}^{Cyl} = \frac{16(1-\Pi)^2}{\pi^2} \cdot 1.86 \cdot \lambda_{SM} \cdot A \cdot p^{1/3}$$

$$= 3.02\ (1-\Pi)^{4/3} \cdot \left[\frac{(1-\mu^2)}{Y} \cdot p\right]^{1/3} \cdot \lambda_{SM} \; . \tag{2.34}$$

For spherical particles we have

$$\lambda_{Cont}^{Spheres} = 3.7\ \frac{(1-\Pi)^2 \cdot \lambda_{SM}}{r} \cdot a$$

$$= 3.44\ (1-\Pi)^{4/3} \cdot \left[\frac{(1-\mu)^2}{Y} \cdot p\right]^{1/3} \cdot \lambda_{SM} \; . \tag{2.35}$$

In addition, comparison of (2.34 and 35) shows that a smaller λ_{Cont} can be achieved with fibres arranged in a regular lattice than with a bulk of spheres, note that $1/[4(1-\Pi)] \geq 0$.

In both cases, λ_{Cont} is proportional to $p^{1/3}$ because ot the assumption of elastic deformations at the contact area. If the deformations were plastic, an exponent 1/2 would be the consequence [Fried 1969].

A correlation between λ_{Cont} and density ρ of aerogel and perlite powders is reported by Kaganer [1969a p.86]. The experimental data are in rough agreement with (2.35) assuming constant p.

2.3.3 Temperature Dependence of Contact Conductivity

We are now in a position to discuss the temperature dependence of λ_S (still without the inclusion of possible surface contaminations). We will start with the temperature dependence of λ_{SM}. As for heat flow through gases, there is enough literature on solid-state physics that treats solid thermal conductivity at any desired level. We will again confine ourselves to comments on the temperature dependence of the relevant parameters.

Temperature Dependence of Solid Thermal Conductivity in Crystalline and Amorphous Substances

A first insight into the problem of which parameters and intrinsic states of a solid influence the solid thermal conductivity λ_{SM} can be gained if we consider four rules for establishing high λ_{SM} in nonmetallic crystals. These rules are derived [Slack 1973] from a theoretical expression [Leibfried and Schlömann 1954] for λ_{SM} in face-centered cubic crystals for the temperature range $T \geq \Theta_D$ (Θ_D denotes Debye temperature)

$$\lambda_{SM} = B \cdot \frac{m \cdot \delta \cdot \Theta_D{}^3}{T \cdot \gamma^2} \ . \tag{2.36}$$

In this expression, B is a constant, m denotes average mass of an atom in the crystal, δ^3 is the average volume occupied by one atom of the crystal, and γ denotes Grüneisen's constant. The four rules are: 1) low atomic mass, 2) strong interatomic bonding, 3) simple crystal structure, and 4) low anharmonicity. Conditions 1 and 2 mean high Θ_D, so that the numerator in (2.36) is maximized for light mass, covalently (i.e., strongly) bonded crystals like diamond or diamond-like binary compounds as BeO or cubic BN, because $\Theta_D{}^3$ dominates. Condition 3 means low number N of atoms per crystallographic primitive unit cell. If N is large, optical phonon branches with small group velocity contribute very little to λ_{SM} (see the literature cited by Slack that supports this expectation). Condition 4 means small γ. Most crystals have $1 \leq \gamma \leq 2$. The Grüneisen parameter is smallest for diamond-like compounds, i.e. in crystals with the highly regular (simple) tetrahedral coordination.

Conversely, the smallest λ_{SM} which are desirable e.g. for thermal insulations can be found if these rules are violated.

It is well known that λ_{SM} does not only depend on the degree of (geometrical) symmetry of a crystal lattice (the smaller the lattice symmetry, the lower is λ_{SM}). In addition, masses of ions that are strongly different, for instance in TiO_2 or YrO_2 instead of BeO or MgO, or that exhibit complex atomic arrangements such as $3Al_2O_3 \cdot 2SiO_2$ or $MgO \cdot Al_2O_3$ instead of graphite, NaCl, KBr, CaO, usually favour small λ_{SM} (Fig.2.6) [Kingery et al. 1958]. Clearly these criteria should be considered when thermal insulations are developed. However, aside from "static" geometrical or chemical anisotropy and lattice defects, thermoelastic waves, i.e. the heat motion of the

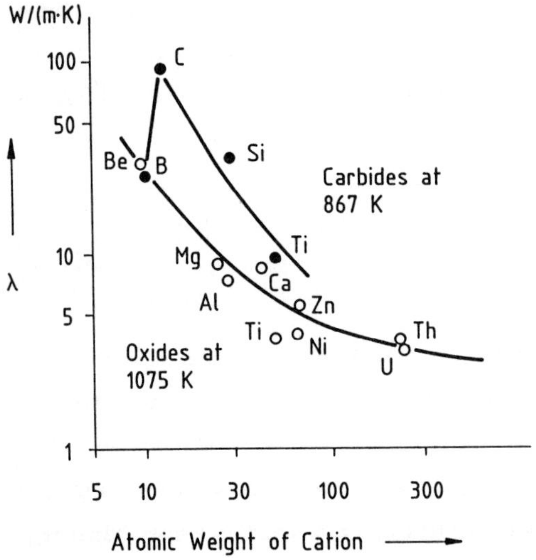

Fig.2.6. Total thermal conductivity $\lambda = \lambda_{SM}$ of some oxides and carbides plotted versus atomic weight of cation [Kingery et al. 1958]

lattice itself, are the reason for anharmonic perturbations that increase strongly with temperature. The probability for the scattering of elastic waves of thermal frequency is again described by the inverse value of a mean free path ℓ. In solids, $1/\ell$ is composed of contributions from various scattering processes [Klemens 1983]

$$1/\ell = 1/\ell_b + 1/\ell_p + 1/\ell_i \tag{2.37}$$

where ℓ_b is the mean free path due to a limitation by grain boundaries or macroscopic inclusions, ℓ_p arises from point defects, and ℓ_i describes the above mentioned anharmonic interactions between the thermoelastic waves themselves (it is assumed that aside from the most important U-processes, i.e. phonon-phonon backscattering, absorption of phonons caused by electron transitions, phonon Raman scattering, excitation of spin waves, are all incorporated in ℓ_i). All components in (2.37) are a function of frequency ω. The thermal conductivity that is caused by any mobile carrier of an excitation obeys the general law

$$\lambda_j = (1/3)\, C_{V,j} \cdot \rho_j \cdot \ell_j \cdot v_j \tag{2.38a}$$

which is reflected for instance by (2.3).

Accordingly, the lattice component of the thermal conductivity (using $C_V = c_V \cdot \rho$) reads

$$\lambda_L = (1/3) \int C_V(\omega) \cdot \ell(\omega) \cdot v \cdot d\omega \quad . \tag{2.38b}$$

In this equation, v is the average phonon or sound velocity and $C_V(\omega) \cdot d\omega$ describes contributions to specific heat by acoustic or mobile lattice modes within the frequency interval $d\omega$.

In metals, an expression like (2.38a) that takes into account the electronic component λ_{El} of the thermal conductivity has to be considered instead of - or in addition to - (2.38b). The spread of spin and exciton waves may additionally contribute to total λ_{SM}.

In (2.37) the intrinsic mean free path ℓ_i varies with the inverse of temperature, $\ell_i \propto 1/T$ [Klemens 1969], provided T is much larger than the Debye temperature Θ_D (at low and high temperatures, significant deviations from this dependence occur [Klemens 1983]). The number of point defects is considered to be independent of temperature. This is certainly true at moderate temperatures; however, disturbances of translation symmetry increase in the case of strong thermal excitations. At low and medium temperatures, the component ℓ_p of ℓ is thus limited by interactions of high frequency phonons with a nearly constant number of disturbances, e.g. $1/\ell_p = A\cdot\omega^4$ [Klemens 1955], where the coefficient A is proportional to defect concentration. Thus ℓ_p is also independent of temperature in this temperature range. If the temperature goes to zero, high-frequency phonons freeze successively. Only acoustic, i.e. long-wavelength trains remain that are hardly scattered by the point-like imperfections (the analogy to Rayleigh and long-wavelength radiation scattering is obvious). Accordingly, ℓ_p increases with decreasing T. A possible approximation is $\ell_p = \text{const}+B/T^2$.

Figure 2.7a illustrates the well-known temperature dependence of ℓ in a nonmetal and reveals how its components ℓ_b and ℓ_i manifest themselves at certain temperatures (we consider a very pure substance): at low T, ℓ grows exponentially with decreasing T until it is much larger than the lattice constant. It will be restricted by surface scattering as soon as it is comparable to grain diameter d. For $T \to 0$, ℓ eventually becomes constant: $\ell = C\cdot d$ (C=const). In this case, the temperature dependence of λ_{SM} is given only by the temperature dependence of the specific heat C_V, which is proportional to T^3 as long as $T \ll \Theta_D$ (Fig.2.7b). As a consequence, $\lambda_{SM} \propto T^3$ for $T \ll \Theta_D$. To give an example, λ_{SM} for LiF illustrates this dependence in Fig.2.7c. If T increases, ℓ_i decreases according to $\ell_i \propto \exp[\Theta_D/(g\cdot T)]$ (where $g \simeq 2$) because phonon-phonon scattering processes (U-processes) increase. This is reflected in Fig.2.7c by the corresponding decrease of λ_{SM} at $T \geq 10$ K. At $T \gg \Theta_D/g$, C_V is nearly constant according to the Dulong-Petit law. Since $\ell_i \propto 1/T$, λ_{SM} decreases with 1/T.

The interatomic separation is a lower limit for ℓ_i. Therefore, the thermal conductivity finally reaches a constant value and no longer decreases with increasing T.

A strong restriction of ℓ to about 4 nm at $T \geq 400$ K coinciding approximately with the diameter of SiO_4 tetrahedra is responsible

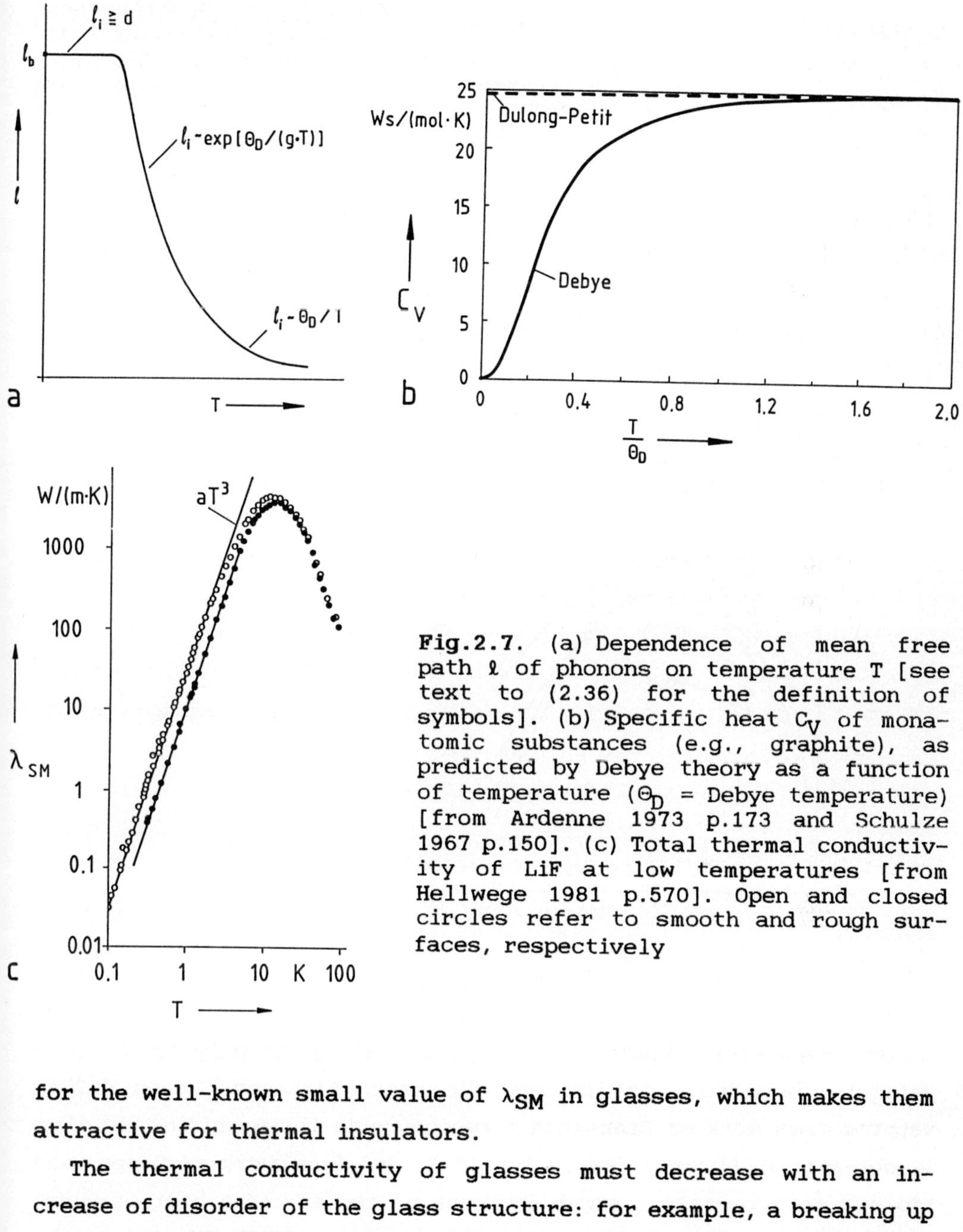

Fig.2.7. (a) Dependence of mean free path ℓ of phonons on temperature T [see text to (2.36) for the definition of symbols]. (b) Specific heat C_V of monatomic substances (e.g., graphite), as predicted by Debye theory as a function of temperature (Θ_D = Debye temperature) [from Ardenne 1973 p.173 and Schulze 1967 p.150]. (c) Total thermal conductivity of LiF at low temperatures [from Hellwege 1981 p.570]. Open and closed circles refer to smooth and rough surfaces, respectively

for the well-known small value of λ_{SM} in glasses, which makes them attractive for thermal insulators.

The thermal conductivity of glasses must decrease with an increase of disorder of the glass structure: for example, a breaking up of the silicon-oxygen network occurs by heavy doping of the glass with alkali or alkaline earth oxides as more tretrahedra are bonded to only two other tetrahedra [Ammar et al. 1982].

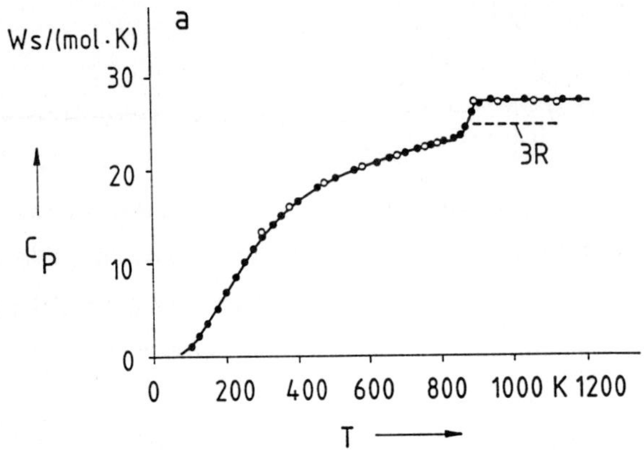

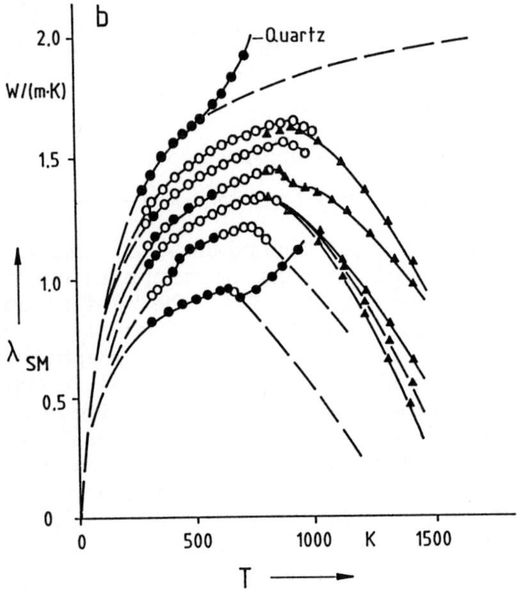

Fig.2.8. (a) Specific heat C_P of boron-aluminum-silicate glass (Duran 50) plotted versus temperature [Coenen 1974]. (b) Total thermal conductivity of different glasses plotted versus temperature [Coenen 1974]

The temperature dependence of λ_{SM} is again given only by the temperature dependence of the specific heat. Although Coenen [1974] reports from work of Dransfeld a resonance in the specific heat that is observed in Pyrex glass at T = 10 K, and from work of Cooper and his own measurements a steep increase of the specific heat of Duran 50 glass due to a configuration slip at $T \geq 800$ K (Fig.2.8a), the specific heat follows the Debye curve, at least between these limits. This behaviour of the specific heat with T is typical for all glasses. As a consequence, the λ_{SM} of glasses and other *amorphous* solids increases with T within this interval (whereas the λ_{SM} of *crystal* substances decreases with increasing T if $T >> \Theta_D/g$). However, at $T \geq$ 800 K, all glasses exhibit a pronounced decrease of λ_{SM} (Fig.2.8b) [Coenen 1974].

It is not a contradiction to these general laws that Espe [1960 p.105] reports a relation $\lambda_{SM} \propto a \cdot T^x$ (x<1) for quartz at temperatures below 200°C although this temperature is clearly below Θ_D/g. Also, the increase of λ_{SM} of clear molten quartz with T follows a weak parabolic law [Espe 1960 p.462].

Experimental values of thermal conductivity λ_{SM} are documented for ceramics and glasses in numerous data tables [Espe 1960; Goldsmith et al. 1961; Landolt-Börnstein 1972].

Temperature Dependence of Thermal Conductivity of Liquids

Although it is not the subject of this review to deal with heat transport phenomena in liquids, a few remarks on the thermal conductivity seem appropriate.

The temperature dependence of thermal conductivity λ_L of *liquids* is seen from Bridgman's model [Bridgman 1923]

$$\lambda_L = 2k \cdot v \cdot D^{-2} \; . \tag{2.39}$$

Here k denotes the Boltzmann constant, and D a mean distance of molecules in a cubic array. The sound velocity is given by v. To give an example, D and v of H_2O or acetone at room temperature amount to 0.31 nm and 1485 m/s or 0.5 nm and 1190 m/s, respectively [Jakob 1964 p.81]. In organic liquids, v decreases and D increases if the temperature increases. As a consequence, λ_L of organic liquids decreases with increasing T. Below 110°C, v of H_2O increases more quickly than D with increasing T, so that λ_L increases with increasing T. However, at elevated temperatures, λ_L decreases again. The behaviour of λ_L of H_2O is rather exceptional.

Temperature Dependence of Young's Modulus of Glasses and Ceramics

Whereas numerous data are available for λ_{SM} at a variety of temperatures, experimental values for modulus of elasticity Y and Poisson's ratio μ of ceramics and glasses, which both appear in (2.25,30 and 32) are rather scarce. Espe [1960] listed some Y and μ values determined at room temperature; Kruse et al. [1971] reported values for a few optical materials; and Ardenne [1973] presented data for fibrous substances. The magnitude of Y for different glasses at room temperature amounts to $5\cdot10^{10}$ to $8\cdot10^{10}$ N/m^2.

A possible temperature dependence of Y of glasses can be estimated as follows. Espe [1960 p.76] reported on a method to determine the viscosity η by the elongation of a glass filament

$$\eta = C_1 \cdot \frac{G\cdot\ell}{A\cdot\Delta\ell/\Delta t} . \tag{2.40}$$

In this equation, C_1 is a constant, A is the cross section of the filament, and G denotes a weight that elongates the filament of length ℓ within a time interval Δt by an amount $\Delta\ell$. We can use this relation to couple η to Y: if the time interval Δt is constant, the relative change $\Delta\ell/\ell$ is given by Hooke's law, viz. $G/A = Y\cdot\Delta\ell/\ell$ or

$$\eta = C_2\cdot Y , \tag{2.41}$$

where C_2 is another constant. The temperature dependence of η of different glasses is well known in the range $400°C \le T \le 800°C$ [Espe 1960 p.82]: $\eta = C_3\cdot\exp(C_4/T)$. Thus we obtain within the range of validity of Hooke's law

$$Y = C_5\cdot\exp(C_6/T) , \tag{2.42a}$$

where $C_3 \ldots C_6$ are constants. Experimental data for Y at different temperatures [Fahrenkrog 1983] have been used to determine the constants C_5 and C_6 and to calculate Y(T) for glass as given in Figure 2.9a. This relation will be used below to calculate the temperature dependence of λ_S of glass fibres and spheres.

The value of Y for ceramics is clearly higher than that of glass. As in the case of amorphous substances, little is known about its temperature dependence. The corresponding value for Al_2O_3 at room temperature amounts to $35\cdot10^{10}$ N/m^2 [Espe 1960 p.612]. This value can be achieved at high baking temperatures. It decreases by about

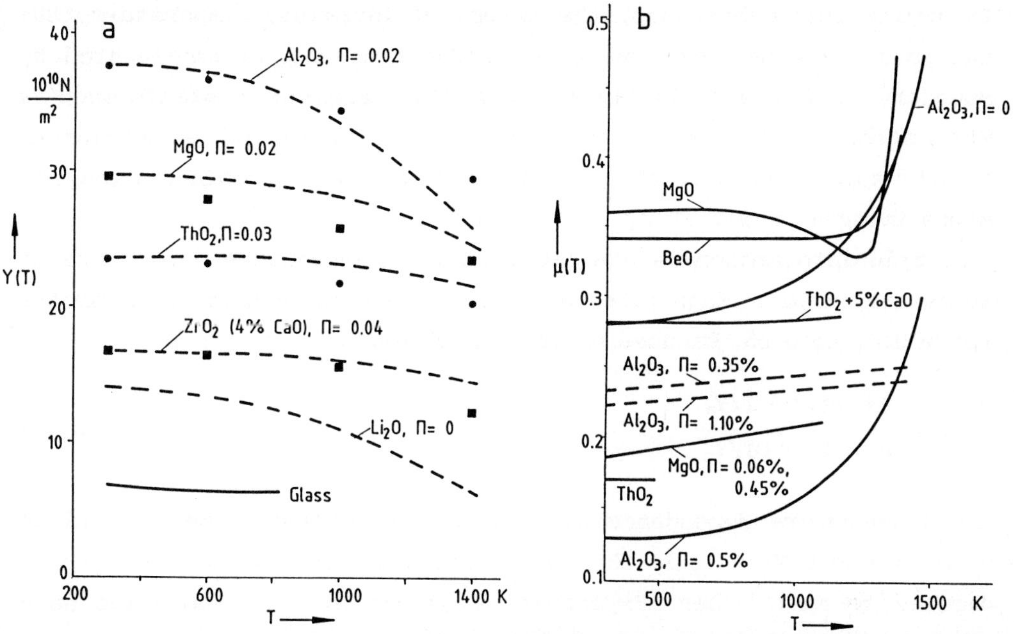

Fig.2.9. (a) Young's modulus of elasticity of glass (solid curve, from (2.42a) using $C_5 = 5.86 \cdot 10^{10}$ N/m^2 and $C_6 = 41.03$ K) and of ceramics as a function of temperature T and for different porosities Π. Data points (full circles) and dashed theoretical curves are from Krikorian [1985] (work performed under auspices of DOE at Univ. of California, Lawrence Livermore Laboratory). (b) Poisson's ratio μ for several oxides as a function of temperature T and for different porosities Π; from Krikorian [1985] (work performed under auspices of DOE at Univ. of California, Lawrence Livermore Laboratory)

one third at T = 1250°C. The corresponding decrease in Y for BeO is even stronger and amounts to about two thirds. According to Krikorian [1985] the temperature dependence of Y for ceramics should be given by an expression that is similar to (2.42a):

$$Y = Y(0) - \left(\frac{T}{T_{Melt}}\right) \cdot Y(0) \cdot \exp\left(1 - \frac{T_{Melt}}{T}\right) , \qquad (2.42b)$$

where T_{Melt} is the melting point (T_{Melt} has been correlated with Y in this reference). Y(0) is the value of Y(T) at absolute zero [possibly (2.42a) should be reformulated in an equivalent manner]. Krikorian reports good agreement when comparing (2.42b) with experimental data for Al_2O_3, MgO, ThO_2, UO_2 and MgO- and CaO-stabilized ZrO_2 (Fig.2.9a).

In crystalline substances, the effect of incrasing temperature thus causes a slow decrease of Y until the decrease is accelerated by boundary sliding and finally melting. Since elastic constants such as Y(T) strongly depend on the internal structure of a substance, (2.42b) applies only if no phase transitions occur within the temperature interval under study.

Little information is available on the temperature dependence of Poisson's ratio μ. Magnitudes of μ for glasses and ceramics lie in the range 0.22 to 0.25. Because of its definition

$$\mu = \frac{\Delta \ell_1/\ell_1}{\Delta \ell_2/\ell_2} \propto \frac{1/Y_1}{1/Y_2} \tag{2.43}$$

the temperature dependence of μ in amorphous substances should be small because $Y_1 \propto Y_2$ in the two directions 1 and 2 that are perpendicular to each other. Experimental values of μ for ceramics have been collected by Krikorian [1985] and are plotted versus temperature in Fig.2.9b. Below 1000 K, the temperature dependence of μ of these substances is small.

Conclusion for Temperature Dependence of Contact Conductivity

We are now ready to calculate $\lambda_{Cont}(T)$, at least of spherical glass particles and glass fibres, according to (2.24,25,29,30) (Fig.2.10). The slope of λ_{Cont} is almost linear between $400\ K \leq T \leq 600\ K$.

The temperature dependence of $\lambda_{Cont} = \lambda_{SC}$ and of $\lambda_{S,0}$, see (2.33), can thus be used to solve the equation of conservation of energy (1.15). On this basis, we can analyze local temperature distributions and local thermal conductivities, at least within certain temperature intervals, if a possible influence of adsorbed gas layers can be neglected. This problem will be discussed now.

Influence of Adsorbed Gases on Contact Conductivity

A contribution to the total thermal conductivity resulting from adsorbed gas layers can arise from two different heat transport mechanisms. First, adsorbed gases can intensify existing point contacts between solid particles: if the vapour pressure is sufficiently

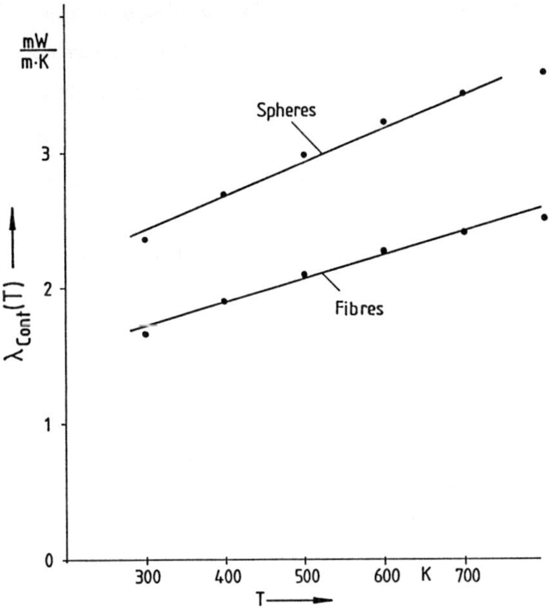

Fig.2.10. Contact thermal conductivity $\lambda_{Cont}(T)$ of spherical and fibrous glass particles, from (2.24,25,29), as a function of temperature (Π = 0.9, pressure load p = 9.81 · 10^4 N/m^2, μ = 0.22, Y(T) of Fig.2.9a and experimental $\lambda_{SM}(T)$ of borosilicate glass taken from Espe [1960 p.105]). Solid curves are linear least squares fits $\lambda_{Cont} = a_1 + a_2 \cdot T$ to the calculated λ_{Cont} (symbols) yielding a_1 = 1.21 mW/(m·K), a_2 = 0.00169 mW/(m·K^2) for fibres and a_1 = 1.70 mW/(m·K), a_2 = 0.00244 mW/(m·K^2) for spheres

high, pores and necks in the contact area will be successively filled by capillary condensation, and (2.24 and 29) no longer correctly describe λ_{Cont}. Second, it is well known that mass flow through a porous medium increases if physical sorption is possible.

Two-dimensional convection in movable adsorbed films that is accelerated by local temperature gradients can exert a very strong influence on total λ: Little et al. [1962] report clearly detectable alterations in measured heat transfer through graphite powder if only 5% of the total particle surface is covered by adsorbed Ar. The rate of two-dimensional mass flow can even be larger than a volume flow through pores. Mass flow is generally accompanied by heat flow. At low temperature and low gas pressure, two-dimensional convection can thus dominate all other heat transfer mechanisms and lead to variations of λ with T that little resemble the λ of a dry substance. An investigation of the necessarily small radiative component is too uncertain in these cases. We will thus confine the remaining part of this subsection to the discussion of possible alterations of the *contact* resistance and their influence on heat transfer.

According to Madhusudana and Fletcher [1981] a contribution to $\lambda_{S,0}$ by adsorbed gases is of minor importance if the gas pressure is below 100 Pa.

Kaganer [1969a p.8] reported on investigations made by Bogomolov[5] on the influence of adsorbed gas layers on the two-phase thermal conductivity of a dispersed system. According to Bogomolov's model, the thickness of an adsorbed gas layer could be twice as large as the particle radius in the case of large porosity (this estimate is realistic only if the vapour pressure is sufficiently high). If this estimate is applied to fumed silica (diameter of single particles between 5 and 10 nm), 50 monolayers of H_2O molecules would be necessary to account for the thickness of the adsorbed gas. This is possible only if the vapour pressure of H_2O is near 100%. Most dispersed media exhibit a considerably lower vapour pressure. As a rule, experimental work described in Chaps.4 and 5 has been performed either in vacuum or with dry substances.

According to Veale [1972 p.68], silica particles are still covered by about eight OH-groups per $(nm)^2$ after deposited H_2O is desorbed from the powder. OH-groups are responsible for the formation of the well-known chain and skeleton structures that are observed for instance in aerogels or fumed silica (Fig.2.11). Below 400°C there is no irreversible destruction of hydrogen bonds between single silica particles. If both the H_2O vapour pressure and the mechanical load are low, a contact between neighbouring silica particles is thus of a purely molecular, i.e. van der Waals' nature[6] [i.e., expressions for λ_{Cont} as given in (2.24,29) are not applicable to this problem; rather, the term $\lambda_{S,O}$ could account for these contributions to λ_S].

Single H_2O molecules may be deposited on the silica particle surface in free positions that are located between the indented silanol groups within the particle-particle contact area. It is not known to what extent these molecules contribute to contact heat transfer, that is, how they alter $\lambda_{S,O}$. Only at elevated temperatures ($T \leq$ 400°C) are these molecules irreversibly removed, so that the contact conductivity again reduces to the proper $\lambda_{S,O}$.

[5]As was the case with other papers cited by Kaganer, the Russian original was not available.

[6] Electrostatic forces are important only for the *formation* of chains or skeletons; charged particles create chains, electrically neutral particles build up spherical clusters; these forces play only an insignificant role with regard to mechanical strength of a bulk [Rumpf 1961 p.394].

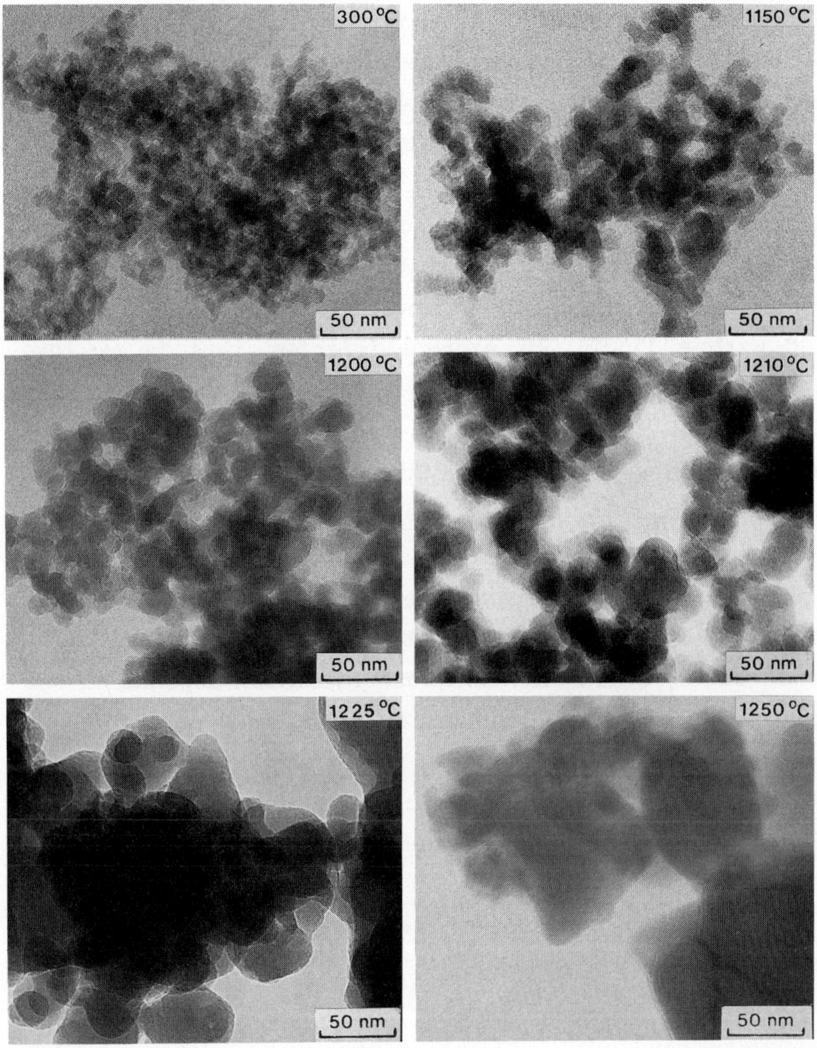

Fig.2.11. Transmission electron micrographs of slices of SiO_2 aerogel treated at various temperatures. Note the increase of particle volume at elevated temperatures. From [Mulder and Van Lierop 1986]

Outside the actual particle-particle contact area, several H_2O *layers* may accumulate. If the separation d of particle surface within the necks exceeds about 2 nm and if the vapour pressure is sufficiently high, capillary condensation can occur. This leads to an increase of the proper contact area. If d is of the order of the diameter of *macrocapillaries* ($\geq$0.1 μm), the heat transfer problem can be solved by a "macroscopic" model [Jackson and Black 1983; Bjurström et al. 1984] that explicitly treats the content of water, ethylene glycol, kerosene and other liquids as a macroscopic parameter in the calculation of the total thermal resistance. For such a model, the mean free path for example of H_2O molecules must be small compared to the capillary radius. Jackson and Black [1983] find that the greatest increase of effective thermal conductivity occurs in the meniscus structure regime where the granular particles become thermally coupled with a liquid. In *microcapillaries,* that is at locations close to the proper contact, heat transfer obeys a law of $\lambda(T^n,p)$. Here the mean free path of H_2O molecules is limited by the capillary radius [Luikov 1966 p.225]. The vapour pressure p depends on the radius of curvature of the capillary.

Compared with fumed silica, distances between particle surface separations of glass fibres within and near the contact area are considerably larger. In this case, the contacts are constituted not by molecular bonds but by solid bridges whose formation depends on surface geometry and roughness, mechanical load, and mechanical or thermal stresses. Figure 2.12a,b shows that surface roughness of a borosilicate glass fibre may limit the closest distance between two fibre surfaces to about 50 to 80 nm (like the curvature of the necks the surface roughness thus also limits the mean free path ℓ_{Gas} of the gas molecules within the contact area; $\ell_{Gas} \simeq 60$ nm at atmospheric pressure and room temperature).[7]

In summary, it is not possible to give a quantitative relation that explicitly accounts for local adsorbtion phenomena within or close to the contact area and their influence on $\lambda_{Cont}(T)$. Only on a macroscopic scale can λ_{Cont} be represented as a function of H_2O content [Luikov 1966 pp.270-276]. Bjurström et al. [1984] extended a model originally derived [Luikov et al. 1968] by inclusion of adsorbed substances. Experimental values of total λ were reproduced without accounting for radiative contributions, however. As a consequence, no

---

[7] An effective gap width d is discussed by Madhusudana and Fletcher [1981]. The gap width d depends not only on surface roughness h [Rapier et al. 1963] but also on the accommodation effect: $d = h+2g$, where g is the temperature jump distance.

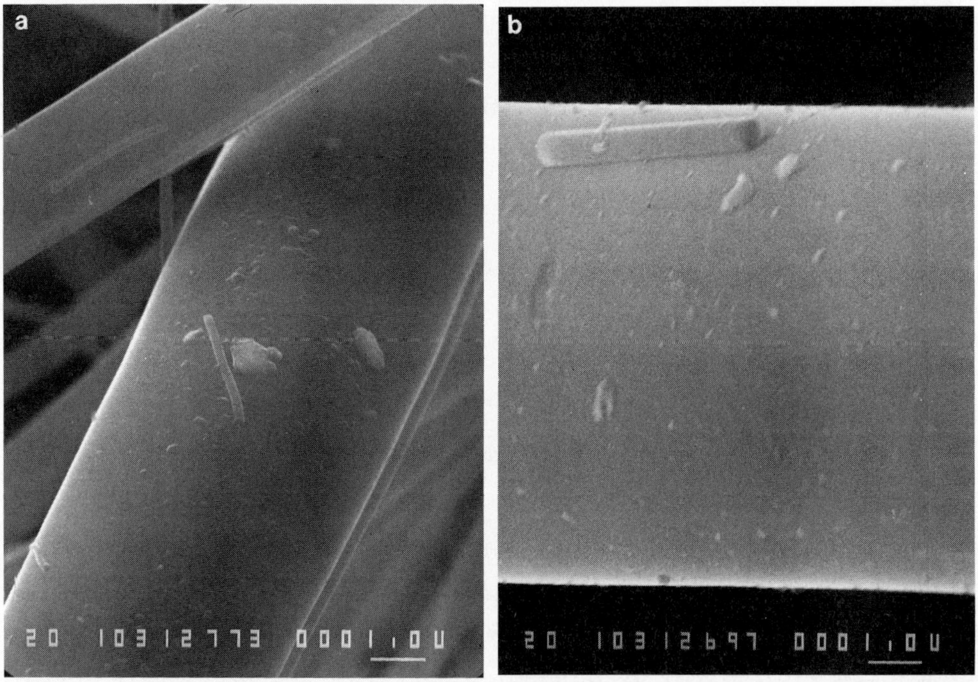

Fig.2.12a,b. Scanning electron micrographs of a single borosilicate glass fibre showing surface roughness. The figures indicate that glass fibres have no micropores, i.e. the first term in (2.8) cancels if λ_{Gas} is calculated as in Fig.2.2b. The length of the horizontal bars indicates 1 μm

conclusion on the temperature dependence of an alteration of $\lambda_{S,0}$ by adsorbed gas layers can be drawn from that work. The same argument applies to that of Jackson and Black [1983]: although good agreement is reported from a comparison of predictions obtained with a unit cell model and experimental data for the total λ of sand, only approximate conclusions on the temperature dependence of $\lambda_{S,0}$ can be drawn if the temperature dependence of the conductivity of the liquid forming the meniscus is known. The total λ of sand increases almost linearly with the λ of the liquid.

2.4 Radiative Flow

2.4.1 Survey: How Do Single Radiation-Particle Interactions Enter the Equation of Transfer?

The equation of radiative transfer (1.2) reads in more detail [Siegel and Howell 1972 p.689]

$$\mu \cdot \frac{di_\Lambda'(\tau_\Lambda)}{d\tau_\Lambda} = - i_\Lambda'(\tau_\Lambda) + (1 - \Omega_\Lambda) \cdot i_{\Lambda b}'(\tau_\Lambda) + \frac{\Omega_\Lambda}{4\pi} \cdot \int i_\Lambda'(\tau_\Lambda, \omega_i) \cdot \Phi_\Lambda(\omega, \omega_i) \cdot d\omega_i \tag{2.44}$$

if a plane, absorbing, emitting and scattering medium is considered. In this equation, μ denotes the cosine of the angle β that is defined in Fig.2.13 between the normal of the (infinitely extended) plane and an arbitrary direction of a beam. Aside from the previously introduced spectral quantities $i_\Lambda'(\tau_\Lambda)$, τ_Λ and Ω_Λ (Sects.1.3,6), $\Phi_\Lambda(\omega,\omega_i)$ denotes the scattering phase function that depends on the solid angles ω and ω_i. The solid angle ω is defined with regard to radiation that is scattered from all inhomogeneities contained in a volume element, whereas ω_i refers to incident radiation. Furthermore, $i_{\Lambda b}'(\tau_\Lambda)$ denotes spectral intensity of black body radiation that is emitted at the position τ_Λ.

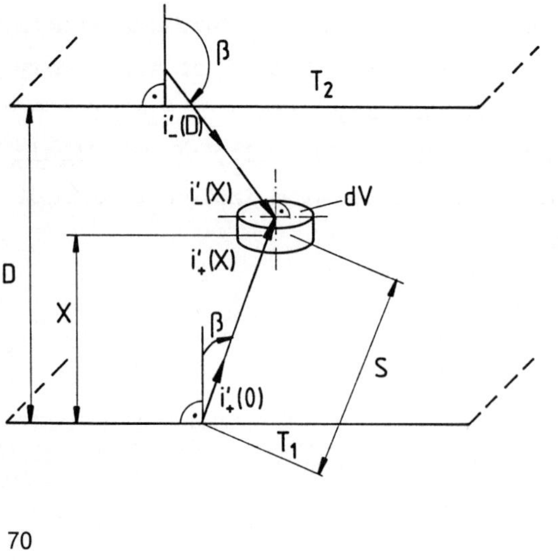

Fig.2.13. Definition of coordinates and directional intensities in a plane layer. After [Siegel and Howell 1972]

The intensity i_Λ' that is the solution of (2.44) is, in general, not isotropic: $i_\Lambda' = i_\Lambda'(\tau_\Lambda,\mu)$. Isotropic *intensity* $i_\Lambda'(\tau_\Lambda)$ (not isotropic *scattering)* is, as we will see in Chap.3, a consequence of nontransparency. Approximate methods to solve (2.44), for instance the well-known (classical) two-flux model, assume isotropic intensity for simplification.

In order to solve (2.44) and (1.15) we need, aside from the parametric functions discussed in the previous sections, the parametric functions E_Λ, Ω_Λ and $\Phi_\Lambda(\omega,\omega_i)$.

Let us for the moment assume that each inhomogeneity within the volume element V (e.g., each particle embedded in a matrix that has a different refractive index) contributes to absorption and scattering of radiation as if its neighbours were not present. In short, we assume "independent scattering". In this case, the extinction coefficient E_Λ of a medium and extinction cross section $C_{Ext,\Lambda}$ of an inhomogeneity are simply coupled by the number N per unit volume of inhomogeneities if all $C_{Ext,\Lambda}$ are equal

$$E_\Lambda = (N/V)\cdot C^{Sph}_{Ext,\Lambda} = \frac{(3\pi/2)\cdot(\rho/\rho_0)\cdot Q_{Ext,\Lambda}}{\Lambda\cdot x}, \tag{2.45a}$$

$$E_\Lambda = (N/V)\cdot C^{Cyl}_{Ext,\Lambda} = \frac{4(\rho/\rho_0)\cdot Q_{Ext,\Lambda}}{\Lambda\cdot x} \tag{2.45b}$$

in the case of spherical or cylindrical particles, respectively. [The $Q_{Ext,\Lambda}$ denotes relative extinction cross sections, see (2.52); x is the scattering parameter, see (2.57), and the ρ,ρ_0 denote densities of the dispersed medium and the solid substance, respectively]. For the solution of (2.44) it is thus necessary to find all $C_{Ext,\Lambda}$ [in addition to Ω_Λ and $\Phi_\Lambda(\omega,\omega_i)$]. In order to solve (1.15) we also need the temperature dependence of these functions.

Regardless of whether the medium is a continuum or not, calculation of the required functions $C_{Ext,\Lambda}$, Ω_Λ and $\Phi_\Lambda(\omega,\omega_i)$ is straightforward, provided we are able to correctly describe the scattering of a plane wave by a single inhomogeneity. We have to find solutions of Maxwell's equations at each boundary separating two microscopic regions of space with different refractive indices, and to calculate the electromagnetic fields in the interior of an inhomogeneity and also the scattered fields. Mie was the first to succeed in calculating intensity distributions of light scattered by colloidal solutions of very small spherical metal particles [Mie 1908]. The rigorous Mie scattering theory is described in various standard textbooks which, in part, contain solutions of the problem for a variety of refractive indices, spherical and cylindrical particles of different diameters, wavelengths and internal particle structures, for instance for particles that consist of subshells of differing composition [Stratton 1941; Born and Wolf 1965; Hottel and Sarofim 1967 which is confined to a short representation, Kerker 1969; Wickramasinghe 1972; van de Hulst 1981; Bohren and Huffman 1983]. As a pure continuum theory, Maxwell's equations and Mie's solution are applicable at all wavelengths.

Before high-speed computers made available the fast algorithms and series expansions that are needed for the calculation of the c_{Ext}, scattering amplitudes and phase angles had to be taken from data tables [Lowan et al. 1946]. The recently published book by Bohren and Huffman [1983] contains computer programs for the calculation of the necessary parametric functions, and examples.

What about particles of arbitrary shape or irregular surface? A fractal approach for modeling surface irregularities was recently suggested [Bourrely et al. 1986] for calculation of light scattering by particles that are large compared to the wavelength. Although Mie's solution is strictly speaking applicable only to idealized particle geometries, it is surprising that calculations are in good agreement with experiment, even if the rigorous assumptions on particle geometry or fluctuations in diameter are not fulfilled.

2.4.2 The Rigorous Mie Theory of Scattering

If a scattering particle is composed of a perfect electrically conducting substance, the scattering process reduces to a mere diffraction problem (if the particle is large compared to wavelength, total reflection occurs). There are no internal electromagnetic fields within the particle. If the substance is not a perfect conductor or is a dielectric, however, a complete treatment of the scattering problem requires solutions for incident, internal and scattered waves of a vector wave equation. The vector wave equation follows immediately from Maxwell's equations. In the absence of sources and with the complex refractive indices (index Λ omitted) $m = n-i\cdot k$, we obtain using $\exp(i\omega\cdot t)$ as time dependence of the fields **E** and **H**

$$m^2 = \mu\cdot(\epsilon - i\cdot\frac{4\pi\cdot\sigma_{El}}{\omega}) \qquad \text{or} \tag{2.46a}$$

$$m^2 = \mu\cdot\epsilon = n^2 \tag{2.46b}$$

for an electrical conductor or a dielectric, respectively. Using the wave number $\underline{k} = 2\pi/\Lambda$, the vector wave equation reads

$$\nabla^2\Psi + \hat{k}^2\cdot\Psi = 0 \ . \tag{2.47}$$

In this equation, $\hat{k}^2 = m^2\cdot\underline{k}^2$, μ and ϵ are the magnetic permeability and dielectric constant, ω is the frequency and σ_{El} is the electrical conductivity.

Spherical Particles

Born and Wolf [1965 pp.634-645] showed that the mathematical problem of solving the vector wave equation can be reduced to a search for solutions to the scalar wave equations

$$\nabla^2\ {}^e\Pi + \hat{k}^2\cdot{}^e\Pi = 0\ , \tag{2.48a}$$

$$\nabla^2\ {}^m\Pi + \hat{k}^2\cdot{}^m\Pi = 0\ , \tag{2.48b}$$

provided a set of two independent electromagnetic fields (${}^e\mathbf{E}, {}^e\mathbf{H}$ and ${}^m\mathbf{E}, {}^m\mathbf{H}$) can be derived from the scalar potentials ${}^e\Pi$ and ${}^m\Pi$ that satisfy $\nabla\times\mathbf{E} = -k_2\cdot\mathbf{H}$ and $\nabla\times\mathbf{H} = +k_1\cdot\mathbf{E}$ (we have $k_1\cdot k_2 = -\hat{k}^2$). In turn, it is sufficient to search for solutions ${}^e\Pi$ and ${}^m\Pi$ of (2.48). Boundary conditions for **E** and **H** transform into boundary conditions for the ${}^{e,m}\Pi$. Mie's problem thus reduces to the construction of the

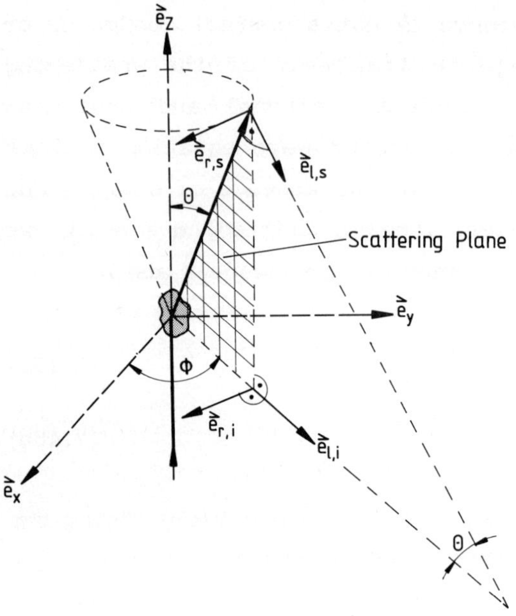

Fig.2.14. Definition of angles θ and ϕ and of incoming (index i) and scattered (index s) components of vector E, parallel (index ℓ) and perpendicular (index r) to plane of scattering of light by spherical particles. After [van de Hulst 1981]

scalar potentials ${}^{e,m}\Pi$ that have to be determined for incident, internal and scattered waves

$$\Pi = R(r)\cdot\Theta(\theta)\cdot\hat{\phi}(\phi) \quad . \tag{2.49}$$

This yields a set of equations that allows determination of scattering coefficients a_ℓ and b_ℓ. These coefficients are used to expand the components $E_{\rho,s}$ and $E_{\theta,s}$ and $E_{\phi,s}$ of the scattered wave (index s) in a spherical coordinate system (r,θ,ϕ) using $\rho=\hat{k}\cdot r$ and Hankel functions of first order and Legendre polynomials. Whereas the radial part $E_{\rho,s}$ quickly goes to zero, $E_{\theta,s}$ and $E_{\phi,s}$ are identified as the components $E_{\ell,s} = E_{\theta,s}$ and $E_{r,s} = -E_{\phi,s}$ that are defined parallel (index ℓ) and perpendicular (index r) to the scattering plane (Figure 2.14).

The scattering matrix S is defined as a 2x2 *amplitude* transformation matrix

$$S = \begin{bmatrix} S_2 & S_3 \\ S_4 & S_1 \end{bmatrix} \tag{2.50}$$

with complex $S_1(\theta),...S_4(\theta)$ that are functions of the scattering angle θ. The S matrix is used to calculate the components of the scattered wave by

$$\begin{bmatrix} E_{\ell,s} \\ E_{r,s} \end{bmatrix} = S \cdot \begin{bmatrix} E_{\ell,i} \\ E_{r,i} \end{bmatrix} \cdot \frac{e^{i\hat{k}\cdot r}}{i\hat{k}\cdot r} \tag{2.51}$$

from the amplitudes of the incident wave (index i). If the particles are strictly spherical, the scattering matrix contains only diagonal elements S_1 and S_2.

Since S_1 and S_2 are complex, scattered radiation is elliptically polarized even if incident radiation is linearly polarized. The phase difference that is needed to describe elliptically polarized radiation follows from Stoke's parameters [van de Hulst 1981 pp.41-42]. For the calculation of Stoke's parameters, elements of an *intensity* transformation matrix F [van de Hulst 1981 pp.44-46] have to be determined. All elements of F are known since all elements of *amplitude* transformation matrix S are already given (the elements of F are measurable quantities; compare Liou et al. [1983]).

If a denotes the radius of a spherical particle, the optical theorem yields the extinction cross section

$$C_{Ext} = (4\pi/\underline{k}^2)\cdot Re\{S(0)\} = \pi\cdot a^2\cdot Q_{Ext} , \tag{2.52}$$

because $S_1(0) = S_2(0) = S(0)$ in this case. It is convenient to use the ratio $Q_{Ext} = C_{Ext}/(\pi\cdot a^2)$ of extinction cross section C_{Ext} and geometrical cross section $\pi\cdot a^2$ of a spherical particle. According to (2.52), extinction by a spherical particle does not depend on polarization of incoming radiation.

The phase function Φ is calculated from

$$\Phi(\theta) = \frac{|S_1(\theta)|^2 + |S_2(\theta)|^2}{\int\left[|S_1(\theta)|^2 + |S_2(\theta)|^2\right]\cdot d\omega} . \tag{2.53}$$

In order to calculate the albedo Ω we finally need an expression for the scattering cross section C_{Sca}, or $Q_{Sca} = C_{Sca}/(\pi \cdot a^2)$. This yields

$$\Omega = \frac{Q_{Sca}}{Q_{Ext}} . \tag{2.54}$$

More explicit expressions for Q_{Ext}, Φ and Ω follow from the series expansions of $S_1(\theta)$ and $S_2(\theta)$ which contain the scattering coefficients a_ℓ and b_ℓ (with all relevant information on the optical properties of a particle contained in the a_ℓ and b_ℓ):

$$Q_{Ext} = \frac{2}{x^2} \cdot \sum_{\ell=1}^{\infty} (2\ell+1) \cdot \mathrm{Re}\{a_\ell + b_\ell\} , \tag{2.55}$$

$$Q_{Sca} = \frac{2}{x^2} \cdot \sum_{\ell=1}^{\infty} (2\ell+1) \cdot (|a_\ell|^2 + |b_\ell|^2) . \tag{2.56}$$

In (2.55,56) x denotes the scattering parameter

$$x = \underline{k} \cdot a = \pi \cdot d/\Lambda \tag{2.57}$$

where d is the particle diameter. The value of the wavelength Λ has to be taken at positions outside of the particle.

Figure 2.15 shows Q_{Ext} for dielectric spherical particles of different refractive indices. The resonance-like maxima of Q_{Ext} at certain values x have been used in the literature to define an "optimum particle diameter" d_{Opt} to achieve large extinction coefficients E in thermal insulation development [Kaganer 1969a p.41]. A similar consideration is used to optimize back scattering cross sections of pigment particles [Jaenicke 1956]. For illustration, Figure 2.16 shows d_{Opt} for dielectric spherical particles of different refractive indices as a function of wavelength.

Note that this procedure gives an optimum extinction coefficient E_Λ (at fixed optimum diameter d_{Opt}) only for *one* wavelength. Furthermore, nothing has been said about a possible anisotropy of scattered radiation. Strong forward scattering can completely annihilate an apparently optimum extinction performance of a particle

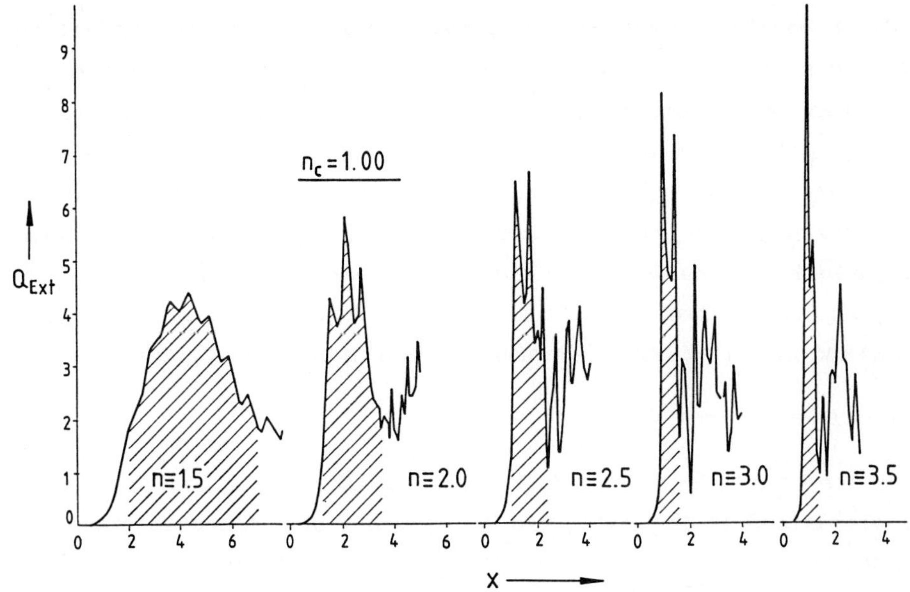

Fig.2.15. Relative extinction cross sections Q_{Ext} of spherical dielectric particles of different refractive indices n calculated from rigorous Mie theory (n_C denotes the refractive index of the continuum that surrounds the particles). Curves are given as a function of the scattering parameter $x = \pi \cdot d/\Lambda$, see (2.57)

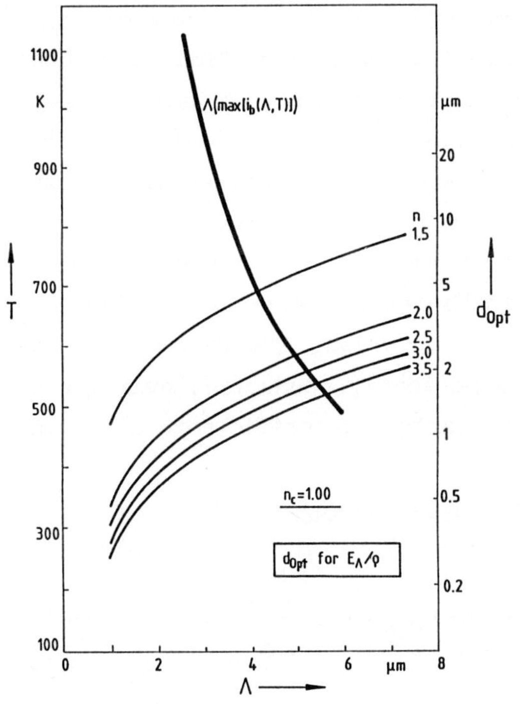

Fig.2.16. Optimum particle diameters for creation of maximum E_Λ/ρ of spherical dielectric particles following from rigorous Mie theory as a function of wavelength Λ given for different (real) refractive indices n. The d_{Opt} are defined with respect to the maximum values of Q_{Ext} in Fig.2.15. The thick solid curve denotes Wien's displacement law (i.e. for T = 900 K radiation temperature, n=2, we have $d_{Opt} \simeq 1.5\ \mu m$)

(Chap.4). In addition, dependent scattering may call for a correction of the simple formulae (2.45a,b). Hottel et al. [1971] consider this problem when optimizing a TiO_2 pigment. Finally, the observed resonances in Q_{Ext}/x are in all real cases considerably broadened because the particle diameter fluctuates in a bulk. To see the extent to which resonances and ripple structure in the Q_{Ext} disappear when particle radii fluctuations increase see an investigation by Yurevich and Konyukh [1975] that assumes a gamma-distribution of radii, or the comments and calculations of Bohren and Huffman [1983 pp.296-299].

Cylindrical Particles

Calculations of Q_{Ext}, Φ and Ω of fibre-like particles are enormously simplified if the ratio ℓ/d of length ℓ of a fibre to diameter d is very large (for glass fibres, $\ell/d \geq 1000$ in most cases). For perpendicular incidence, the calculation of the internal and scattered fields is very similar to that of spherical particles: the field components are again found as derivatives of two scalar potentials. These potentials are solutions to a scalar wave equation such as (2.48), which is now formulated in cylindrical coordinates. However, two possibilities A and B for orientation of the fibre symmetry axis z with regard to the field vectors of incoming radiation have to be considered separately (Fig.2.17). As a consequence, we have two sets of scattering coefficients: $a_{\ell(A)}, a_{\ell(B)}$ and $b_{\ell(A)}, b_{\ell(B)}$. Like the scattering matrix S for spherical particles, the scattering matrix T for cylindrical particles is a 2x2 array of amplitudes $T_1 \ldots T_4$. For perpendicular incidence ($\phi = 0$) the normal to the scattered wave is perpendicular to the z axis, i.e. the scattered wave is a cylindrical wave (if $\phi \neq 0$, a conical wavefront is observed). The matrix elements T_3 and T_4 are zero if $\phi=0$ or if a perfectly conducting cylinder is considered (in the latter case the scattering coefficients $a_{\ell(A)}$ and $b_{\ell(B)}$ disappear because $m \to \infty$).

The corresponding expressions for Q_{Ext} and Q_{Sca} read

$$Q_{Ext(A)} = (2/x) \cdot \mathrm{Re}\{T_1(0)\} = (2/x) \cdot \mathrm{Re}\{b_{0(A)} + 2 \sum_{\ell=1}^{\infty} b_{\ell(A)}\} , \qquad (2.58)$$

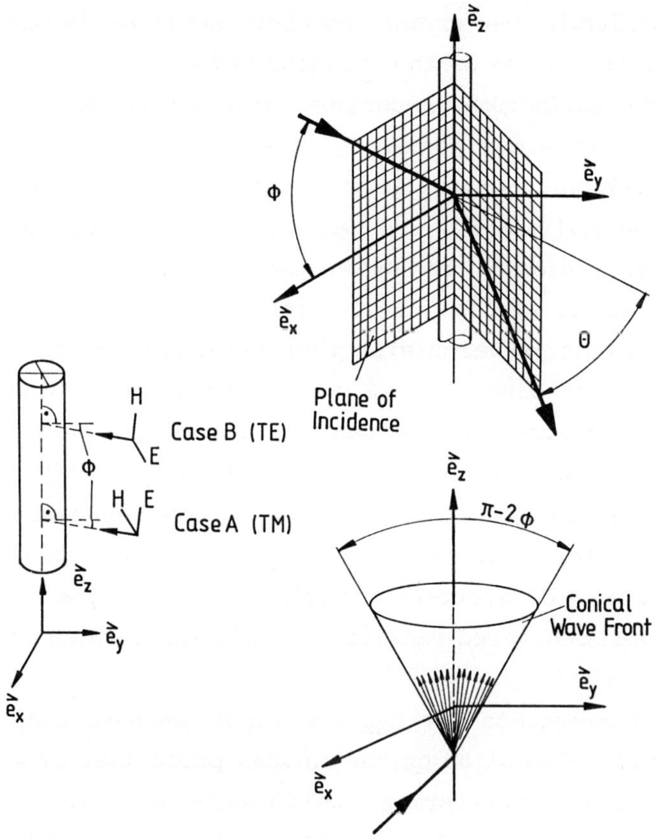

Fig.2.17. Definition of angles θ and ϕ and of TM or TE modes of scattering of light by cylindrical particles [after Kerker 1969, and Bohren and Huffman 1983]. Reprinted by permission of John Wiley & Sons, Inc.

$$Q_{Ext(B)} = (2/x)\cdot \mathrm{Re}\{T_2(0)\} = (2/x)\cdot \mathrm{Re}\{a_{0(B)} + 2\sum_{\ell=1}^{\infty} a_{\ell(B)}\} , \tag{2.59}$$

$$Q_{Sca(A)} = (2/x)\cdot[|b_{0(A)}|^2 + 2\sum_{\ell=1}^{\infty}(|b_{\ell(A)}|^2 + |a_{\ell(A)}|^2)] , \tag{2.60}$$

$$Q_{Sca(B)} = (2/x)\cdot[|a_{0(B)}|^2 + 2\sum_{\ell=1}^{\infty}(|a_{\ell(B)}|^2 + |b_{\ell(B)}|^2)] . \tag{2.61}$$

As with spherical particles, the a_ℓ and b_ℓ contain the complete information on the optical properties of the cylindrical particle. From

(2.58,59) it is again concluded that extinction cross sections do not depend on the state of polarization of the incoming radiation.

When investigating the *extinction* properties of spherical or cylindrical particles, it is thus not necessary to know the polarization of the incoming radiation. However, if the *scattering* cross sections (and phase function and albedo) are calculated, knowledge of polarization of incoming radiation is indispensable. As a consequence, the same requirement applies if an "effective" extinction cross section (or coefficient) is calculated that takes into account a possible anisotropy of scattering (Sects.3.3.3 and 3.4.3). According to Hottel et al. [1970] radiative transfer can be treated with sufficient accuracy without inclusion of polarization effects if multiple scattering is observed (i.e. optical thickness $\tau_0 > 1$). However, radiation scattered from very thin, strongly reflecting metallic fibres is completely polarized at all scattering angles. Since T_3 and T_4 disappear, all incoming and scattered radiation has the *same* state of polarization.

By inspection of the expressions for Q_{Ext}, Φ and Ω, we have seen that the relevant physical information on the optical properties of a particle is contained in the scattering coefficients a_ℓ and b_ℓ [summations over ℓ in (2.55,56) and (2.58-61) occur for purely mathematical reasons]. As a consequence, an "intrinsic" temperature dependence of E, Φ and Ω may arise only from a corresponding dependence of the refractive index m (all quantities entering the expressions for a_ℓ and b_ℓ are functions of m, particle diameter d and wavelength Λ; if d and Λ are fixed, uncertainties in a_ℓ and b_ℓ are exclusively due to uncertainties in m). We therefore have to look for reliable sources of and possible uncertainties in m (Sect.2.4.5).

2.4.3 Dependent Scattering

Application of (2.45a,b) is valid only if dependent scattering is negligible. At least two criteria have been proposed in the literature to decide whether dependent scattering is present or not: one of them denies dependent scattering if the cross section (in set theory notation) of $C_{Ext,1}$ and $C_{Ext,2}$ of two particles 1 and 2 is zero. A second criterion makes a decision on the basis of the ratio of particle clearance and wavelength (see below).

The first criterion was suggested by Weber [1960]. Figure 2.18 gives the minimum (center to center)[8] distance D between spherical dielectric particles (for large values of x) that is necessary for vanishing dependent scattering (the D in this figure refers to the resonances of Q_{Ext}; see Fig.2.15). For instance, with a particle diameter d = 5 μm, the minimum distance required to avoid dependent scattering from a "shadow effect" would be $\geq$ 12 μm. Jaenicke [1956] called for a ratio D/d = 3.6 when investigating backscattering of white pigments, which is only in rough accordance with Fig.2.18; the value of n for pigments is usually between 2 and 3. In dispersed opacified powder insulations, distances between scattering opacifiers are on the average indeed larger than 12 μm: if we use 15 weight percent TiO_2 powder with 4 μm grain diameter as an opacifying additive to fumed silica, and if the specific density of the mixture is 300 kg/m^3, we have D = 4d = 16 μm in the idealized situation of a cubic lattice. Since single particles of fumed silica (as the major component of the insulation) hardly contribute to scattering, and because Q_{Ext} does not depend on the polarization of the incoming radiation, dispersed powder insulations are a suitable candidate (among others)

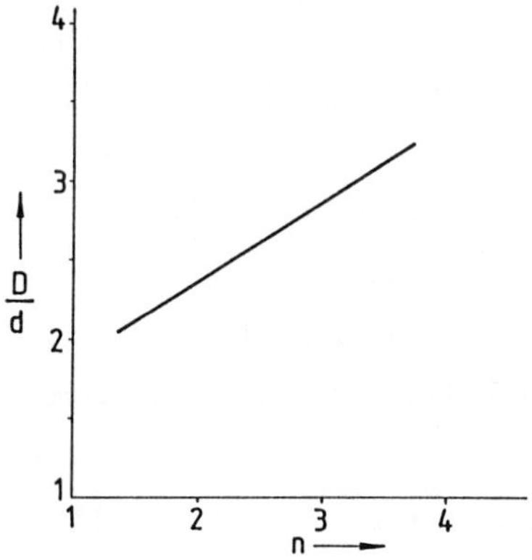

Fig.2.18. Minimum center to center distance D between spherical dielectric particles for establishing independent scattering (d: particle diameter) versus refractive index n

[8] For very small spherical particles and constant Λ, x $\rightarrow$ 0, and there is only absorption. Since Q_{Ext} is very small, dependent scattering for instance in fumed silica is thus unlikely to manifest itself in the extinction coefficient.

to check the predictions of Mie theory in the infrared wavelength region (Sects.4.3,4).

Dielectric or conducting cylindrical particles yield $Q_{Ext} \leq 8$ if $1.4 \leq n \leq 2.5$ and $k \leq 2$ ($m = n - i \cdot k$). If the separation of fibres is at least 8 diameters, dependent scattering should vanish if the first criterion still holds. However, within the contact area between two fibres, the cross section of $C_{Ext,1}$ and $C_{Ext,2}$ certainly does not vanish. In a regular lattice of fibres, the distance ℓ between contacts is given by Kaganer [1969a p.23] as

$$\ell = \frac{\pi \cdot d}{8(1 - \Pi)} \tag{2.62}$$

with Π denoting the porosity. Thus $\ell/d \geq 8$ if $\Pi \geq 0.95$.

We may treat this problem by analogy to the corrections that are necessary in the case of short cylinders. Let $R = \ell/d$. Diffraction theory couples ℓ/d to the ratio [Bohren and Huffman 1983 p.211]

$$\frac{\ell^2}{d^2} = R^2 = \frac{P_e(\theta,i)}{P_e(\theta,ii)} \; . \tag{2.63}$$

In this equation, $P_e(\theta,i)$ and $P_e(\theta,ii)$ are the envelopes of the phase function Φ in directions perpendicular and parallel to the cylinder axis. If $R \geq 10$ there is accordingly little radiation scattered in directions other than those of the cylindrical outgoing wave. In this case, the cylinder may be regarded as infinitely long, i.e. the disturbances from contact area can only be weak.

So far, the simple criterion that uses vanishing cross sections (set theory notation) between $C_{Ext,i}$ and $C_{Ext,j}$ of two particles i and j in a bulk is not selfcontradictory, even when contacting cylindrical particles are considered, provided the distance between contacts is not too small. Since this criterion is entirely based on critical particle separations, it should be possible to deduce from this a "critical porosity" Π' that indicates the onset of dependent scattering if $\Pi \leq \Pi'$. Brewster [1981] reviewed these critical porosities and found that they do not yield an unambiguous separation between the two scattering regimes. We will concentrate briefly on some experiments that lead us to Brewster's improved solution.

Wang et al. [1981] measure extinction cross sections and scattered radiation of two spherical particles not only as a function of distance D but also for different angles α of orientations of the second particle with reference to an incoming beam. Dependent scattering (i.e. alterations to the single particle C_{Ext}) is observed if D < 10d and $\alpha<60^0$. If $\alpha=0$, for example, C_{Ext} decreases by a factor 2, if the distance $\underline{k}\cdot D = (2\pi/\Lambda)\cdot D$ is reduced by a factor 3 (n = 1.363, x = 4.678). Experimental results for C_{Ext} and $i'(90^0)$ are confirmed by calculations of Kattawar and Dean [1983]. Note that instead of relative particle distance D/d, the ratio D/Λ (apart from a constant) is treated as a critical parameter. A very similar procedure has already been suggested by Hottel et al. [1970,1971] and Brewster and Tien [1982b]: it has been experimentally verified that the ratio $(D-d)/\Lambda$ of particle clearance to wavelength indicates which scattering regime is present. If $(D-d)/\Lambda<0.3$, dependent scattering is dominant (Fig.2.19). This critical value can be compared with the investigation of Kattawar and Dean [1983]: using x = 0.9283 we have $d/\Lambda=0.296$ or $\underline{k}\cdot D \le 3.74$ for the onset of dependent scattering. This estimate is confirmed by Kattawar's results: for $\underline{k}\cdot D \le 4$ (n = 1.54) considerable deviations from the undisturbed C_{Ext} are observed for $C_{Ext,r}$ and

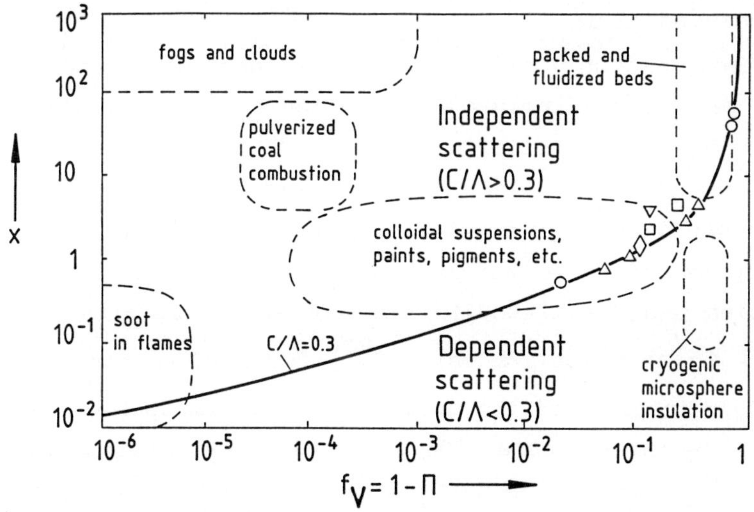

Fig.2.19. Separation of independent and dependent scattering regimes: scattering parameter $x = \pi\cdot d/\Lambda$ versus particle volume fraction $f_V = 1-\Pi$ (Π: porosity). From [Brewster and Tien 1982b]. C denotes particle clearance, C = D-d

$C_{Ext,\ell}$ at $\alpha=0°$ and $90°$ (the indices r and ℓ refer to r- and ℓ-polarized radiation as defined in Fig.2.14). Agreement has thus been found between the work of Wang et al. [1981], Brewster [1981], Brewster and Tien [1982b], and Kattawar and Dean [1983]. Obviously, the simple D/d criterion, as given by Fig.2.18, is too weak; if $x = 0.9283$ we have from $(D-d)/\Lambda<0.3$ an estimate $D/d < 2.01$ instead of $D/d < 2.15$ that follows from Fig.2.18 for $n = 1.54$, i.e. dependent scattering starts *before* extinction cross sections $C_{Ext,1}$ and $C_{Ext,2}$ shade each other.

In a very recent publication [Drolen et al. 1987], the $(D-d)/\Lambda$-limit has been extended to 0.5 instead of 0.3 by introducing as a criterion a 5% deviation from scattering efficiencies calculated using Mie's theory. On this basis, from scattering experiments performed with very small latex particles ($x = 0.205$) and by comparison with predictions of the Percus-Yevick hard sphere model applied to the case $x \to 0$ (see below), it was found that below $1-\Pi = 0.0064$ scattering is presumably independent regardless of the size of the scattering parameter. Like in other scattering experiments with clouds of very small particles, however, this conclusion depends on the assumption that the latex particles (mean particle diameter about 0.03 μm) were not agglomerated. Formation of particle pairs would already be sufficient to shift the scattering parameter to about twice the assumed value.

These approaches yield only criteria for the *onset* of dependent scattering. More generalized studies are directed towards calculating the effect of dependent scattering on the extinction cross section Q_{Ext} in packed-sphere systems. These studies involve the form factor concept known from X-ray scattering theory (a short yet comprehensive description of these models has only very recently been published by Drolen and Tien [1987]). This concept correlates the reduction of Q_{Ext} to an effective value Q_{Ext}^{Eff} for a given particle with the solid angle integral of the product of the form factor $F(\theta)$ (that accounts for the interference effects) and the scattering phase function

$$\frac{Q_{Ext}^{Eff}}{Q_{Ext}} = \frac{1}{4\pi} \int F(\theta)\cdot\Phi(\theta)\cdot d\omega \ . \qquad (2.64a)$$

The form factor $F(\theta)$ depends on a pair distribution function that represents the probability of finding a neighbouring particle at a given distance. The analysis made by Drolen and Tien shows that among different approaches to $F(\theta)$, the Percus-Yevick hard sphere model [Percus and Yevick 1958, Wertheim 1963] is the only one that behaves reasonably at small porosities. For $x \to 0$, this model yields (assuming isotropic scattering) [Twersky 1975]

$$\frac{Q_{Ext}^{Eff}}{Q_{Ext}} = \Pi^4 \cdot \left[1 + 2(1 - \Pi)\right]^{-2} . \tag{2.64b}$$

If the porosity Π equals 0.95, the extinction cross section is reduced to about 2/3 of its proper value. Agreement is found when comparing the Percus-Yevick approach with experiments performed for spherical particles with $x = 0.205$ [Drolen et al. 1987], and $x = 0.529$ and 3.518 ($n = 1.19$) [Drolen and Tien 1987].

Caps [1985] measures specific extinction coefficients of fine TiO_2 powder and glass fibres as a function of density ρ or applied external pressure load, respectively. For TiO_2 an exponential decay of scattering coefficient S_Λ with increasing ρ is reported at wavelengths of 4 and 5 μm that is considerably stronger than the decay predicted by (2.64b). The spectral behaviour of S_Λ at different densities is interpreted as Rayleigh scattering by agglomerates that are formed at different states of compaction.

The depencence of extinction coefficients of glass fibres on ρ as observed by Caps is rather small (Sect.5.4).

2.4.4 Measurements of and Approximations for Single Scattering Phase Functions

Whereas numerous measurements of scattering phase functions Φ at visible wavelengths using spherical or cylindrical particles, hexagonal ice needles, molecule chains, aero- and hydrosols, and water droplets are reported in the literature, experimental determinations of Φ in the infrared region are rather scarce. Most of the work reviewed below uses theoretical results for Φ calculated from Mie theory.

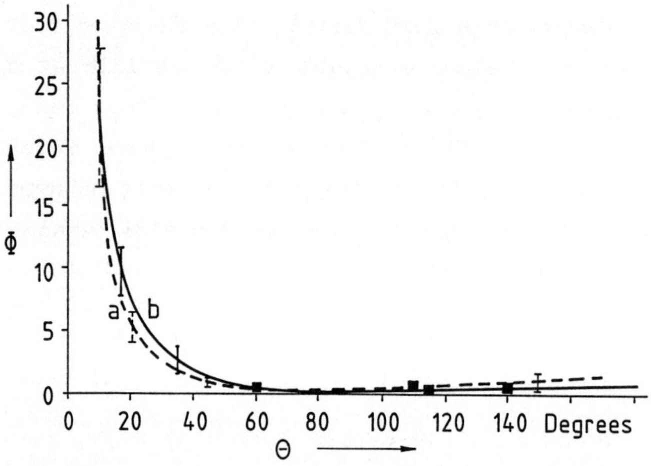

Fig.2.20. Measured scattering phase functions Φ versus scattering angle θ given for polyurethane foam (a) and glass fibres (b). Data are taken at Λ = 9.64 μm and at low density, bars indicate standard deviations [Glicksman et al. 1987]. Reprinted by permission of Pergamon Journals, Inc.

Glicksman et al. [1987] report measured scattering phase functions of polyurethane foam and glass fibres at 9.64 μm wavelength (Fig.2.20). It is obvious that forward scattered radiation in both samples (cell diameter of the foam about 0.5 to 1 mm) is extremely strong, which is clearly the result of diffraction increasing with increasing diameter of the scatterer.

It would be desirable to incorporate extremely anisotropic scattering such as the strong forward scattering shown above into a simple quantity that can be used with sufficient accuracy instead of the exact integral over $i'(\tau,\omega_i)\cdot\Phi(\omega,\omega_i)\cdot d\omega_i$ in (2.44). For this purpose, this subsection discusses briefly approximate expressions for the phase function and the definition of an asymmetry factor μ needed in Chap.3 for the introduction of an "effective extinction coefficient" E that enables us to compare calorimetric, spectroscopic and numerical work.

Many attempts have been made in the literature to approximate scattering phase functions $\Phi(\mu)$ by closed form solutions [van de Hulst (1980 pp.306-307]. Angular distribution of scattered radiation can be extremely complicated (Fig.2.21) so that closed form solutions in many cases yield only a mean degree of anisotropy.

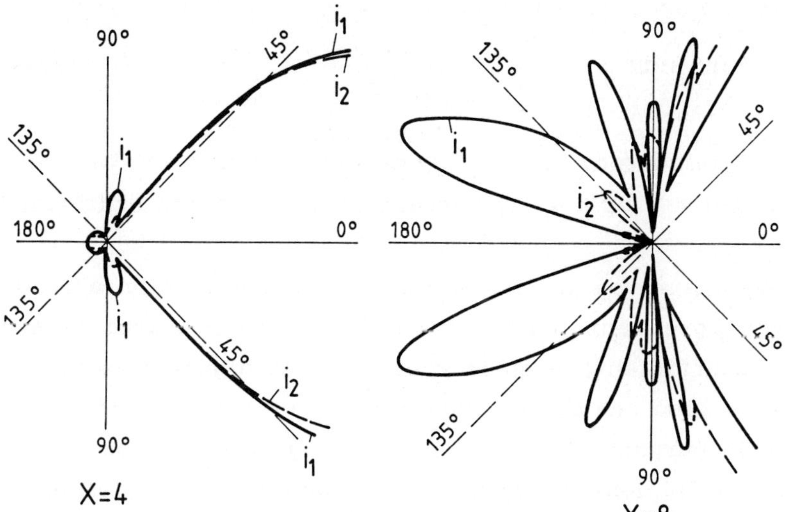

Fig.2.21. Polar diagrams of scattering phase functions for linearly polarized light and spherical dielectric particles (n=1.25) given for $x = \pi \cdot d/\Lambda = 4$ (left) and 8 (right). Reproduced with permission from [M. Born, E. Wolf: *Principles of Optics*, 3rd.ed. (Pergamon, Oxford 1965)]. Angles given in the diagrams denote the scattering angle θ

For an approximation of the phase function $\Phi(\mu)$ we need the albedo Ω and an anisotropy factor $\bar{\mu}$. This factor follows from a series expansion of $\Phi(\mu)$ in Legendre polynomials $P_\ell(\mu)$. If we assume axial symmetry, we have

$$\Omega \cdot \Phi(\mu) = \sum_{\ell=0}^{\infty} \omega_\ell \cdot P_\ell(\mu) \ . \tag{2.65a}$$

The coefficients ω_ℓ are given by

$$\omega_\ell = \frac{2\ell + 1}{2} \cdot \Omega \cdot \int_{-1}^{+1} \Phi(\mu) \cdot P_\ell(\mu) \cdot d\mu \ . \tag{2.65b}$$

If scattering is isotropic, the series in (2.65a) terminates at $\ell=0$: $\omega_0 = \Omega$. If scattering is weakly anisotropic, the expansion can tentatively be terminated at $\ell=1$

$$\omega_1 = (3/2)\Omega \cdot \int_{-1}^{+1} \Phi(\mu) \cdot \mu \cdot d\mu = 3\Omega \cdot \bar{\mu} \qquad \text{where}$$

$$\bar{\mu} = (1/2) \int_{-1}^{+1} \Phi(\mu) \cdot \mu \cdot d\mu \qquad (2.66)$$

is the required anisotropy factor. Complete forward or backward scattering yields $\bar{\mu} = 1$ or $\bar{\mu} = -1$, respectively (isotropic scattering gives $\bar{\mu} = 0$).

It will be demonstrated in Chap.4 that the $\ell=1$ expansion already yields very good agreement between calculated and measured effective extinction coefficients even if scattering is highly anisotropic (this expansion is also one of the fundamentals of the LAS model, see Chap.3). For this purpose, it will be necessary to define an effective extinction coefficient E^* by a transformation $E^* = E \cdot (1-\Omega \cdot \bar{\mu})$ or $\tau_0{}^* = \tau_0 \cdot (1-\Omega \cdot \bar{\mu})$. This transformation lowers E and τ_0 if radiation is scattered into the original direction $\mu = 1$. If the scattering is preferentially backward, the same transformation increases E and τ_0.

Among the approximations to $\Phi(\mu)$ that are listed by van de Hulst [1980], the Henyey-Greenstein phase function, which is well known in astronomy or atmospheric sciences, is given by

$$\Phi(\mu) = \frac{1 - \bar{\mu}^2}{(1 + \bar{\mu}^2 - 2\mu \cdot \bar{\mu})^{3/2}} \; . \qquad (2.67)$$

This function exhibits maxima for $\mu=\pm1$ so that it is suitable for description of radiative transfer not only in its original domain (transparent atmospheres) but has also been applied for transparent liquids [Bergman et al. 1983], for instance. The series expansion of $\Phi(\mu)$ in (2.65) may require more than 100 terms when scattering in aerosols is investigated, and even more than 1000 terms for scattering in clouds [McKellar and Box 1981]. In this case, the extremely narrow peaks of scattered intensity in forward directions are represented by a combination of delta and "residual" Henyey-Greenstein phase functions that have to be normalized properly. A similar, yet more elegant and powerful approach that is especially suitable for solving radiation transfer problems (calculation of radiative fluxes) in the case of highly asymmetric phase functions is given by the δ-M method [Wiscombe 1977].

It will be seen in Sect.3.4.3 that similarity transformations such as $\tau_0 \to \tau_0 \cdot (1-\bar{\mu})$ [van de Hulst and Grossmann 1968] are very important for a simplified calculation of radiative flow by diffusion methods, that is, in nontransparent media.

2.4.5 Refractive Indices

The definition of the real index of refraction, n, follows immediately from the solution of the wave equation that describes the propagation of a plane wave in an isotropic medium of zero electrical conductivity. Using the complete set of Maxwell's equations [Siegel and Howell 1972 p.84] and the same notation as in Sect.2.4.2, we have the wave equation

$$\mu \cdot \epsilon \cdot \frac{\partial^2 E_y}{\partial t^2} = \frac{\partial^2 E_y}{\partial x^2} \qquad \text{with the solution} \qquad (2.68)$$

$$E_y(x,t) = E_{y,0} \cdot \exp\{i \cdot \omega \cdot [t - (\mu \cdot \epsilon)^{1/2} \cdot x]\} \ . \qquad (2.69)$$

Since the dimension of the term $x \cdot (\mu \cdot \epsilon)^{1/2}$ in the exponent of (2.69) must be indentical to the dimension of time t, $(\mu \cdot \epsilon)^{-1/2}$ has the dimension of m/s and is thus identified with the velocity c of electromagnetic radiation in the nonconducting medium

$$c = (\mu \cdot \epsilon)^{-1/2} \ . \qquad (2.70)$$

If we denote by c_0 the velocity of light in vacuum, the ratio $n = c_0/c$ transforms (2.69) into

$$E_y(x,t) = E_{y,0} \cdot \exp\left[i \cdot \omega \cdot (t - \frac{x}{c_0} \cdot n)\right] \ . \qquad (2.71)$$

As a consequence, n is a real number.

If, on the other hand, the isotropic medium has a nonvanishing electrical conductivity, the solution of the wave propagation problem is an exponentially damped plane wave

$$E_y(x,t) = E_{y,0} \cdot \exp\left[i\cdot\omega\cdot(t - n\cdot x/c_0)\right]\cdot\exp(-\omega\cdot k\cdot x/c_0) \ . \qquad (2.72)$$

In this equation, k denotes the attenuation coefficient of the medium. Eq.(2.72) can be rewritten as follows

$$E_y(x,t) = E_{y,0} \cdot \exp\{i\cdot\omega\cdot\left[t - (x/c_0)\cdot(n - i\cdot k)\right]\} \ . \qquad (2.73)$$

Comparison with (2.71) shows that the real refractive index n has been replaced by the complex number $m = n-i\cdot k$.

The three statements in each of the following two groups are therefore equivalent:

1) The medium has zero electrical conductivity; 2) the plane wave propagates in the medium without attenuation; 3) the refractive index is a real number,

or

4) the medium has nonvanishing electrical conductivity; 5) the plane wave is exponentially damped; 6) the refractive index is a complex number.

Statement 2 predicts complete transparency for an ideal, isotropic, nonconducting medium. In reality, complete transparency does not occur because of the excitation of quantum, i.e. discrete energy states that cannot be accounted for by Maxwell's equations, which constitute a pure continuum theory. Furthermore, scattering will be observed if the index of refraction is not isotropic within the medium because of different sorts of perturbations, e.g. if the medium is dispersed. Thus nontransparency is possible with conducting and nonconducting media.

In both cases, the velocity of light in the medium and the wavelenghts of radiation are given by

$$c = \frac{c_0}{n} \quad \text{and} \quad \Lambda = \frac{\Lambda_0}{n} \ , \qquad (2.74a,b)$$

respectively (a measurement of c thus gives no indication of the size of k). By use of these relations, the spectral and total black body emissive power $e_{\Lambda b}$ and e_b transform into $n^2\cdot e_{\Lambda b}$ and $n^2\cdot e_b$, respectively, as given in Sect.3.4.6. Maxwell's equations cannot account for absorption/emission processes that are based on excitation and decay of discrete energy states. As a consequence, the velocity c as given in (2.74a) refers only to that part of the radiation within a medium that does not undergo absorption/emission events. The propagation of absorption/emission events through the medium is a classical diffusion process. The velocity by which the diffusion front propagates through the medium is smaller than the velocity c (elastic scattering is a fast process). Scattered radiation travels with the velocity c in a diffusion-*like* manner through the medium. As a consequence, there are two velocities for the propagation of parts of radiation intensity in an absorbing/emitting and scattering medium. Local intensity distributions and temperature profiles reach static values after time intervals that depend on the ratio of scattering to absorption/emission: The larger the albedo, the more time is required for establishing a static temperature profile if a heat pulse (of appropriate length) was generated within the medium (Fig.3.12).

Fluctuation in particle diameter is a source of error that arises in all applications of Mie theory regardless of the range of wavelengths under study. When dealing with thermal radiation heat transfer, however, two additional difficulties arise: (1) $i_{\Lambda b}'$ (Planck's law) extends to the far infrared, and (2) refractive indices and thus Q_{Ext}, Q_{Sca} and $\Phi(\mu)$ may depend on temperature. Furthermore, dispersed media such as thermal insulations may be composed of a variety of constituents (for instance a "skeleton" of fibres and spherical grains of infrared opacifiers).

Data collections for refractive indices of dielectric materials in the infrared are rather scarce. Most of the older work has been confined to short wavelengths [Landolt-Börnstein 1962 pp.43-404, Waxler and Cleek 1973]. Refractive indices measured at long wavelengths for Si, Ge, MgO, CdS, LiF, NaF, NaCl, KCl, KJ, CsJ and others are given in [Landolt-Börnstein 1962 pp.405-432], for different kinds of glass and SiO_2 in [Wray and Neu 1969, Neuroth 1974, Amrhein 1974, Hsieh and Su 1979] (up to $\Lambda \simeq 1000$ μm), for α-quartz and vitreous silica at 5 μm $\leq \Lambda \leq$ 35 μm in [Gaskell 1976], for fused silica and sapphire at 0.3 μm $\leq \Lambda \leq$ 25 μm [Lang and Wolfe 1983], for Al_2O_3 in 1.5 μm $\leq \Lambda \leq$ 66 μm in [Harris 1955, Harris and Piper 1962], for Al_2O_3 and MgO at $\Lambda \leq$ 10 μm in [Plass 1964], for TiO_2 at $\Lambda \leq 10\ \mu m$ in [Kruse et al. 1971], for Fe_3O_4 up to $\Lambda = 14\ \mu m$ in [Buchenau and Müller 1972], and for polyethylene (Mylar) in $5\ \mu m \leq \Lambda \leq 100\ \mu m$ in [McKay and Timusk 1984]. Dielectric constants of Fe_3O_4 can be found for Λ up to 41 μm in [Schlegel et al. 1979].

The situation is better with metals: Weaver et al. [1981] and Ordal et al. [1983] reported on refractive indices determined at Λ up to several hundred μm. Besides Ordal's tables, Hagen-Rubens' law [Siegel and Howell 1972 p.111]

$$n = k = C_1 \cdot \sqrt{\Lambda/\rho_{El}} \tag{2.75}$$

yields an estimate of n and k at long wavelengths. This law can sometimes be used to complete experimental values by normalizing the constant C_1 in (2.75) to the experimental n or k that have been detected at the largest Λ [Wang and Tien 1983].

Aside from tabulated n and k there are, however, numerous collections of transmission data of inorganic [Nyquist and Kagel 1971] and organic substances and of dielectric constants and dielectric phase angles that can yield qualitative information on $k(\Lambda)$. Also, the extensive data tables for thermal emission coefficients and absorption and reflection properties [Touloukian 1970] are valuable for this purpose, keeping in mind that k of nonconductors is about 10^{-6}, k of absorbing substances amounts to $10^{-3}...1$ and only k of metals reaches values considerably larger than 1. Regarding the value of k, it is a pleasure to recommend [Bohren and Huffman 1983 pp.279-280]; there is also a short survey in [van de Hulst 1981 pp.268-269].

However, the obvious lack of more complete tables of refractive indices at large wavelengths is not yet the proper problem. Rather, it is not at all clear that values of n and k measured at *macroscopic* surfaces (where Fresnel's relations apply) may be used, as is usually done, in calculations of extinction or scattering properties of *microscopic* particles whose geometrical dimensions may be orders of magnitudes below the wavelength. Amorphous silica is suitable to test whether this practice can be justified experimentally since these particles are almost perfectly spherical [Veale 1972 p.41] and completely isotropic. The (macroscopic) refractive index of amorphous SiO_2 is well known over a wide range of wavelengths. Furthermore, particle diameters d are so small (0.04 to 0.1 μm) that in the Rayleigh limit the quantity "C_{Ext}/particle volume" does not depend on d [Bohren and Huffman 1983 pp.309-310]. Finally, agglomeration can be suppressed in vacuum. Bohren and Huffman [1983 p.361] reported satisfactory agreement when comparing calculated and measured C_{Ext}/particle volume for an absorption resonance at Λ=9 μm. Thus ideal particle geometry and isotropic internal structure seem to be the premise for an application of macroscopic refractive indices to interactions of radiation with microscopic particles.

Little is known about the temperature dependence of refractive indices at long wavelengths. For LiF an increase of T leads to a reduction of n at $\Lambda \geq 20$ μm [Landolt-Börnstein 1962 pp.413-414]. Also the dn/dT of NaCl, KCl, KBr, KJ, CsJ, CaF_2 and quartz are negative [Landolt-Börnstein 1962 pp.405-432]. Typical values of |dn/dT| of these substances and of silicate and borosilicate glass [Kunc et al. 1984, Wray and Neu 1969] are between 10^{-6} to 10^{-4} 1/K. Based on the well-known Clausius-Mosotti equation, a temperature dependence of the dielectric constant ϵ can be estimated [Hamann 1983]: since $(\epsilon-1)/(\epsilon+2)$ is proportional to density, it is shown that $(1/\epsilon)\cdot(d\epsilon/dT) \simeq -\beta = (1/V)\cdot(dV/dT)$, taking into account that most dielectrics have $2 \leq \epsilon \leq 3$. This yields $\Delta\epsilon/\epsilon = -10^{-2}$ if $\Delta T = 100$ K. Since $n^2 = \mu\cdot\epsilon$, the theoretical temperature dependence of n is very small. At moderate temperatures the dn/dT are thus too small to cause uncertainties in Q_{Ext}, Q_{Sca} and $\Phi(\mu)$ that could be larger than those arising from fluctuations in particle diameter(Fig. 2.22). A possible temperature dependence of integral extinction coefficients E is in this case restricted to the temperature dependence of the spectral distribution of $i_{\Lambda b}'$, which is used for calculation of E from spectral E_Λ (Sect.3.4.1).

A temperature dependence of refractive index may, however, be important in high temperature experiments. It is well known that thermal emission coefficients ϵ of SiO_2, Al_2O_3, ZrO_2 and MgO decrease between 300 K $\leq T \leq$ 2000 K to $0.2 \leq \epsilon \leq 0.3$, whereas the ϵ of Cr_2O_3 and SiC are almost constant (0.75-0.9 and 0.97, respectively) within

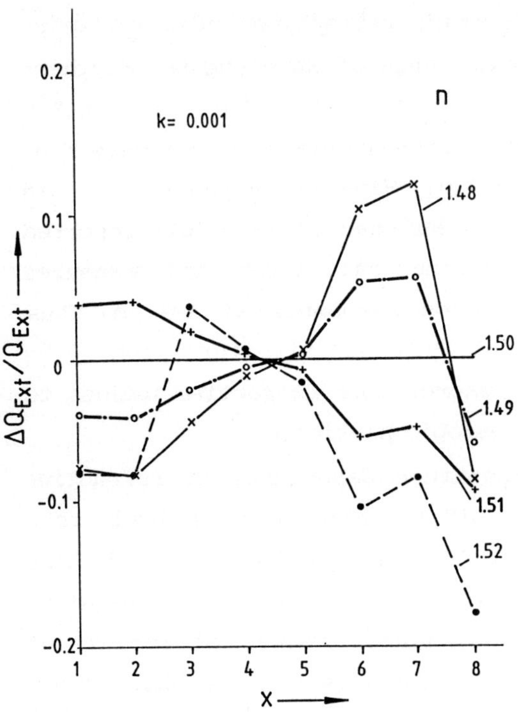

Fig.2.22. Variation $\Delta Q_{Ext}/Q_{Ext}$ of extinction cross section for spherical particles as a function of uncertainty Δn in the refractive index; x denotes the scattering parameter, $x=\pi \cdot d/\Lambda$

this temperature interval [Bauer and Steinhardt 1982]. Older experimental results on spectral dependence (within the range of wavelengths 1 μm $\leq \Lambda \leq$ 10 μm) and temperature dependence of the ϵ of Al_2O_3, MgO, ThO_2, ZrO_2, BeO and CeO_2 at different grain diameters were reported by Ritzow [1934] (the findings of Bauer and Ritzow on $d\epsilon/dT$ do not always agree).

Strong variations with temperature of the refractive index of Al_2O_3,MgO and TiO_2 in the near infrared are also expected from experimental transmission spectra [Cabannes and Billard 1987]: the absorption coefficients A that were calculated from the transmission data by use of a two-flux model expression show a strong increase with temperature at T $\geq$ 300 K (Fig.2.23). Since the measurements were perfomed with nondispersed solids, $E_\Lambda = A_\Lambda$. At $\Lambda \simeq 4$ μm and T $\geq$ 1000 K, A_Λ increases approximately in proportion to T^2 or even T^3. With increasing wavelength, the temperature dependence of A_Λ shifts to an almost linear relationship.

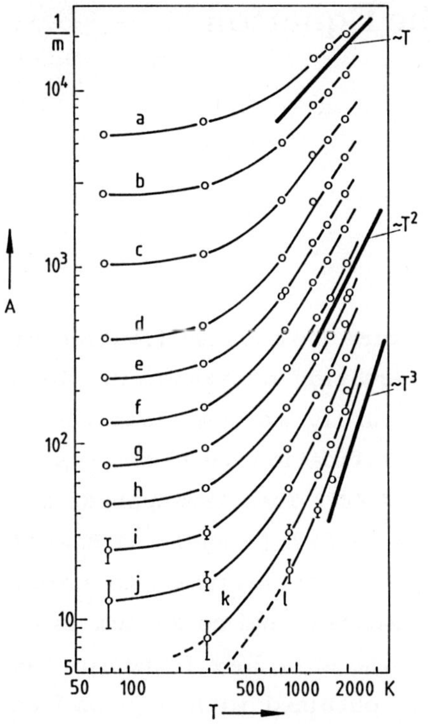

Fig.2.23. Experimental (circles) and calculated (solid curves) temperature dependence of absorption coefficient A of Al_2O_3 in the near infrared [Cabannes and Billard 1987]. The indices a to l refer to the wavelengths 7.1, 6.7, 6.3, 5.9, 5.6, 5.3, 5.0, 4.8, 4.5, 4.3, 4.2, and 4.0 μm, respectively. The thick solid lines are given for an estimate of the temperature dependence of the data

Since $A_\Lambda \propto k_\Lambda$ [Siegel and Howell 1972 p.415] a remarkable dependence of k_Λ on T is obvious for these substances [McLachlan and Meyer 1987].

3. Approximate Solutions of the Equation of Transfer

There is no unique opinion in the literature on a classification of exact or approximate solutions of the equation of transfer: while Viskanta [1982] made his decision on a purely mathematical basis, Siegel and Howell [1972] preferred a classification based on consideration or neglect of physical terms in (2.44). The "transparent gas approximation" or the "cold medium approximation" [Siegel and Howell 1972 pp.462-467] certainly yield only approximate solutions to a real situation, because no gas is totally transparent and no medium is at absolute zero temperature. Nevertheless, the solutions found for the corresponding residuals of (2.44) can be obtained with any desired degree of accuracy, i.e. they are, although incomplete from a purely physical standpoint, mathematically exact. To give another example, Viskanta [1982] treated the discrete ordinate method as an approximate solution. Brewster [1981], on the other hand, used the same method as an exact solution for a control of results achieved with the two-flux model. We will follow in this chapter an older classification of approximate solutions of the complete (2.44) given by Viskanta [1966]. However, it is intended to relate our discussion more closely to calorimetric measurements than has been done in previous reviews. As a consequence, the diffusion model and its modification will occupy most of our attention.

The approximate methods to solve (2.44) that have been used most frequently in the literature to describe radiative transfer in dispersed media can be listed as follows:

a) Two-flux models originally used by Schuster and Schwarzschild, and the very similar Milne-Eddington method;

b) Discrete ordinate methods for cold media that are a generalization of a) because the radiation field is analyzed in more than one direction;

c) Formal solutions of (2.44) for grey, absorbing, emitting, isotropically and anisotropically scattering and heat conducting media;

d) Diffusion methods for radiative transfer in nontransparent heat conducting media;

e) LAS models that use an expansion of the scattering phase function in first-order terms in order to account for anisotropic scattering.

We will not discuss more specialized topics such as the spherical harmonics method[1]. The transparent approximation and the integral kernel substitutions which Viskanta [1966] included in his review will also be omitted.

On the other hand, the "cold medium" approximation will be considered in Sects. 3.1,2 as a favourable case when comparing measured transmission coefficients with calculations.

All media considered below will be assumed to have axially symmetrical optical and thermal properties.

3.1 Applications of the Two-Flux Model

The well-known Schuster-Schwarzschild method [Schuster 1905; Schwarzschild 1906] has become known as the first successful attempt to solve the equation of transfer. Schwarzschild considered the Sun's atmosphere as a grey gas. He found the law of darkening and explained the temperature distribution of the upper atmospheric layer on the basis of a purely radiative equilibrium (i.e., neglecting scattering and heat conduction).

Since all expressions in Sects.3.1,2 apply to an arbitrary wavelength, the index Λ will be omitted, for simplicity. (The medium may, however, be non-grey). The Schuster-Schwarzschild two-flux model assumes that the radiative field can be divided into two oppositely

---

[1] Yurevich and Konyukh [1975] expand directional intensity in terms of Legendre polynomials (moment method). Instead of ordinary Legendre polynomials, application of associated Legendre polynomials leads to the 'spherical harmonics method' [Viskanta 1966 p.206].

oriented components[2] $i_+'(0 \le \mu \le 1)$ and $i_-'(-1 \le \mu \le 0)$, using $\mu = \cos\theta$

$$i_+'(\tau) = \frac{1}{2\pi} \cdot \int_0^{+1} i'(\tau,\mu)\cdot d\mu , \tag{3.1a}$$

$$i_-'(\tau) = \frac{1}{2\pi} \cdot \int_{-1}^{0} i'(\tau,\mu)\cdot d\mu . \tag{3.1b}$$

If an average $\int\mu\cdot i'(\tau,\mu)\cdot d\mu$ of $i'(\tau,\mu)$ is calculated[3], the equations

$$\int_0^{+1} \mu\cdot i'(\tau,\mu)\cdot d\mu = (1/2) \int_0^{+1} i'(\tau,\mu)\cdot d\mu = \pi\cdot i_+'(\tau) , \quad \text{and} \tag{3.2a}$$

$$\int_{-1}^{0} \mu\cdot i'(\tau,\mu)\cdot d\mu = -\ (1/2) \int_{-1}^{0} i'(\tau,\mu)\cdot d\mu = -\ \pi\cdot i_-'(\tau) \tag{3.2b}$$

hold only if $i'(\tau,\mu) = i'(\tau)$, that is if *directional intensity i' is isotropic* (as indicated in Fig.3.1). If, in addition, scattering is isotropic, then $\Phi(\mu) \equiv 1$.

Similarly, integration of (2.44) over $d\mu$ yields two equations

$$\int_0^{+1} \mu\cdot\frac{di'(\tau)}{d\tau}\cdot d\mu + \int_0^{+1} i'(\tau)\cdot d\mu = (1-\Omega)\cdot\int_0^{+1} i_b'(\tau)\cdot d\mu$$

$$+\ \Omega\cdot\int_0^{+1} i'(\tau)\cdot d\mu , \tag{3.3a}$$

---

[2] Since the majority of stars exhibit neither strong rotation nor strong magnetic fields, it is to be expected that they are of spherical shape. Spherical symmetry thus simplifies investigation of transfer problems: it is sufficient to calculate $i'(\tau,\mu)$ parallel to a radius vector.

[3] Ambarzumjan et al. [1957 pp.18–22] reviewed calculation of averages of the equation of transfer.

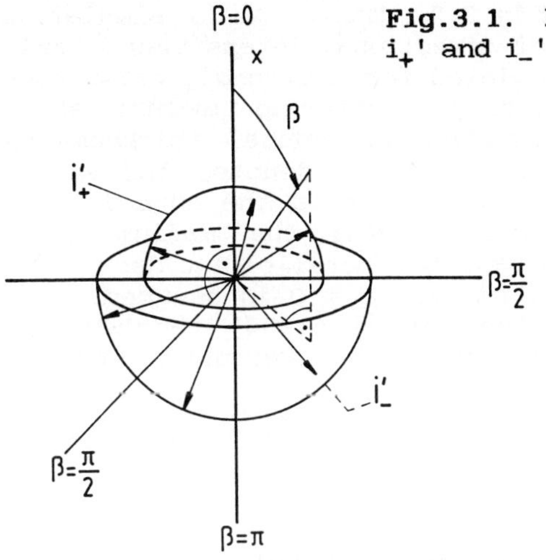

Fig.3.1. Definition of isotropic intensities i_+' and i_-'. After [Siegel and Howell 1972]

$$\int_{-1}^{0} \mu\cdot\frac{di'(\tau)}{d\tau}\cdot d\mu + \int_{-1}^{0} i'(\tau)\cdot d\mu = (1-\Omega)\cdot\int_{-1}^{0} i_b'(\tau)\cdot d\mu + \Omega\cdot\int_{-1}^{0} i'(\tau)\cdot d\mu \tag{3.3b}$$

or using (3.1a,b) and (3.2a,b)

$$\frac{di_+'(\tau)}{2d\tau} + i_+'(\tau) = (1-\Omega)\cdot i_b'(\tau) + \Omega\cdot i_+'(\tau) \ , \tag{3.4a}$$

$$-\frac{di_-'(\tau)}{2d\tau} + i_-'(\tau) = (1-\Omega)\cdot i_b'(\tau) + \Omega\cdot i_-'(\tau) \ . \tag{3.4b}$$

Equations (3.4a,b) are the most frequently used versions of the two-flux model. Isotropic directional intensity and isotropic scattering are necessary conditions for the derivation of these equations. The question arises whether complete fulfillment of both conditions is really necessary. Caps et al. [1984] showed by a Monte Carlo-simulation that also in the case of strong anisotropic scattering or with a strongly anisotropic radiation source, the directional intensity is almost isotropic if $\tau_0 \geq 15$, i.e. after about 15 mean

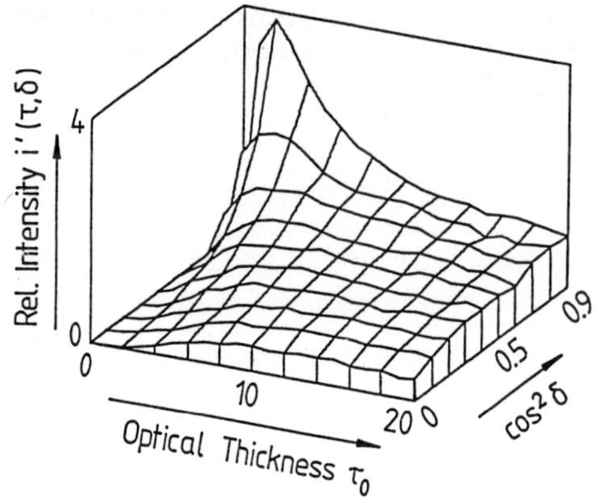

Fig.3.2. Monte Carlo simulation of directional intensities i' calculated for a strongly anisotropically scattering medium as a function of optical thickness τ_0 and $\cos^2\delta$ (δ denotes the angle, with respect to the normal, of a beam leaving the medium at a given τ_0); parallel incident radiation at $\tau_0 = 0$ is assumed. From [Caps et al. 1984] reprinted by permission of Pergamon Journals Ltd.

free paths of photons (Fig.3.2; a similar consideration would show that polarization of incoming radiation also disappears in an optically thick medium of small, randomly oriented, non-metallic particles [Kattawar and Plass 1976]). In turn, it can be concluded that anisotropic directional intensity arises at least in the case of anisotropic scattering if the medium is transparent. Application of (3.4a,b) may then be in error. Thus the two-flux model in its original form applies correctly only to *nontransparent* media [application of (3.4a,b) to transparent media can possibly yield acceptable results if incident radiation is completely diffuse and scattering completely isotropic, see below].

Aside from interstellar dust (which obscures about 85% of the galaxy) protostars are the most important examples for nontransparent media in astrophysics. Protostars are surrounded by an optically thick envelope and can only be detected by being strong sources of infrared radiation, not by high temperature radiation from the core.

Multiplication of (3.4a,b) by the factor π yields, because $\dot{q}_\pm = \pi \cdot i_\pm'$, two expressions for the radiative flux density. In a fruitful extension of (3.4a,b), radiation that is scattered from the flux densities $\dot{q}_\pm$ is accounted for by oppositely oriented multiples $a_\pm \cdot \dot{q}_\pm$ of $\dot{q}_\pm$. For this purpose a backscattering coefficient $b \equiv a_\pm$ is introduced [Wang and Tien 1983; Tong and Tien 1983]:

$$\frac{d\dot{q}_+(\tau)}{d\tau} + 2\dot{q}_+(\tau) = 2(1-\Omega)\cdot\dot{q}_b(\tau) + 2\Omega\cdot(1-b)\cdot\dot{q}_+(\tau) + 2\Omega\cdot b\cdot\dot{q}_-(\tau) \ , \quad (3.5a)$$

$$-\frac{d\dot{q}_-(\tau)}{d\tau} + 2\dot{q}_-(\tau) = +2(1-\Omega)\cdot\dot{q}_b(\tau) + 2\Omega\cdot(1-b)\cdot\dot{q}_-(\tau) + 2\Omega\cdot b\cdot\dot{q}_+(\tau) \quad (3.5b)$$

or using $d\tau = E\cdot dx$, $A = (1-\Omega)\cdot E$, $S = \Omega\cdot E$, in a more familiar form,

$$\frac{d\dot{q}_+}{dx} = -\ 2A\cdot\dot{q}_+ - 2b\cdot S\cdot\dot{q}_+ + 2A\cdot\dot{q}_b + 2b\cdot S\cdot\dot{q}_- \ , \quad (3.6a)$$

$$\frac{d\dot{q}_-}{dx} = 2A\cdot\dot{q}_- + 2b\cdot S\cdot\dot{q}_- - 2A\cdot\dot{q}_b - 2b\cdot S\cdot\dot{q}_+ \ . \quad (3.6b)$$

In these equations, $\dot{q}_b$ equals the hemispherical (spectral) emissive power e_b ($e_{\Lambda b}$) of a black body. The backscattering coefficient b is defined by

$$b = (1/2)\int_{-1}^{0} \Phi(\mu)\cdot d\mu \ . \quad (3.7)$$

For isotropic scattering, $b = 1/2$.

If temperatures are very low, the term $2A\cdot e_b$ in (3.6a,b) can be neglected. For a plane "cold" medium that scatters radiation isotropically, the solutions of (3.6a,b) then read

$$\dot{q}_+ = C_1\cdot\beta\cdot e^{\hat{D}\cdot x} + C_2\cdot e^{-\hat{D}\cdot x} \ , \quad (3.8a)$$

$$\dot{q}_- = C_1\cdot e^{\hat{D}\cdot x} + C_2\cdot\beta\cdot e^{-\hat{D}\cdot x} \quad (3.8b)$$

which can easily be verified (C_1, C_2 and β are constants, and $\hat{D} = 2E\cdot\sqrt{1-\Omega}$). The total $\dot{q}_{Rad}$ is given by $\dot{q}_{Rad} = \dot{q}_+ - \dot{q}_-$.

As is well known, an experimental approach to a cold medium can simply be made by chopping an incident beam at a frequency that is sufficiently high to prevent a significant temperature increase in the sample. A temperature increase would occur if the sample had strong absorption properties.

If in addition heat conduction is negligible, the temperature profile follows from a purely algebraic equation [Siegel and Howell 1972 p.493] if the medium is grey. Let D denote the total thickness of the medium. Transmission and reflection coefficients T(D) and R(D) have to be calculated from

$$T(D) = \frac{\dot{q}_+(D)}{\dot{q}_+(0)} , \tag{3.9}$$

$$R(D) = \frac{\dot{q}_-(0)}{\dot{q}_+(0)} . \tag{3.10}$$

Explicitly, for the cold medium

$$T(D) = \left[(1-\beta^2)\cdot e^{-\hat{D}\cdot D}\right]/(1-\beta^2\cdot e^{-2\hat{D}\cdot D}) , \tag{3.11}$$

$$R(D) = \left[\beta\cdot(1-e^{-2\hat{D}\cdot D})\right]/(1-\beta^2\cdot e^{-2\hat{D}\cdot D}) . \tag{3.12}$$

Equation (3.11) is illustrated in Fig.3.3a. For $\tau_0 = 1$ and $\Omega = 1$ we have T = 1/2. An exact calculation [Hottel 1962] yields T = 0.55. In the case of pure scattering ($\Omega = 1$) or pure absorption ($\Omega = 0$), the transmission coefficients approach the values $T = 1/(1+\tau_0)$ or $T = \exp(-2\tau_0)$, respectively [Kaganer 1969a p.47] for isotropic scattering. Note that in the case of pure absorption the transmitted flux shows a stronger decay than is predicted by Beer's law for directional intensity $[i'(\tau_0)/i'(0) = \exp(-\tau_0)]$.

If R(D) is measured for $D \to \infty$, i.e. with a medium of infinitely large optical thickness, we have from (3.10)

$$R(D\to\infty) = R_\infty = \frac{1 - \sqrt{1-\Omega}}{1 + \sqrt{1-\Omega}} = \beta . \tag{3.13}$$

This is illustrated in Fig.3.3b. As a consequence, Ω can be determined from a reflection measurement. Figure 3.3a shows that measurements of T(D) (with finite D) yield optical thickness τ_0 (or extinction coefficient E). The magnitude of the extinction coefficient E and the contributions by absorption and scattering A and S, respectively, are thus determined in two optical experiments. Applications of this

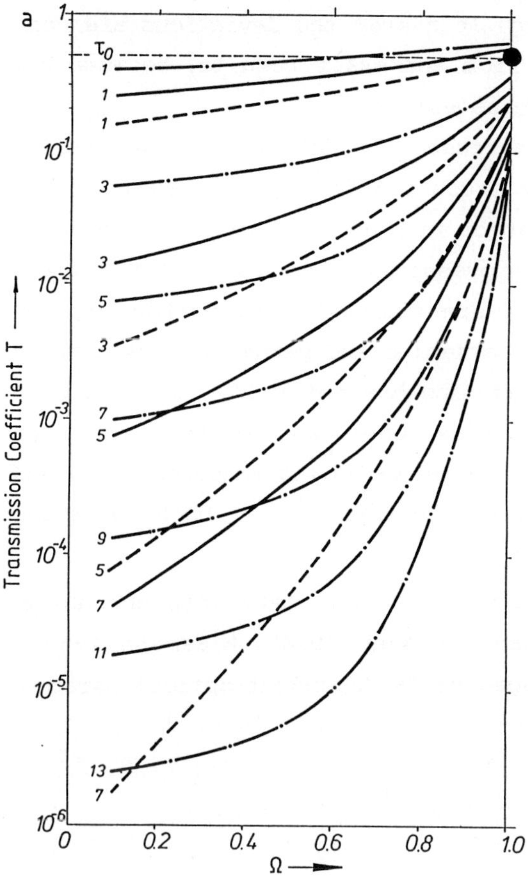

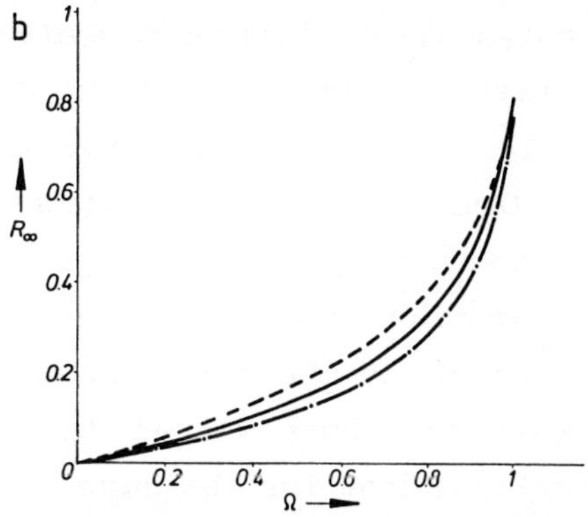

Fig.3.3. (a) Transmission coefficients T given as a function of albedo Ω and optical thickness τ_0 calculated from two-flux model [(3.11), dashed curves] and discrete ordinate method three-flux approximation for diffuse (solid curve) and parallel (dashed-dotted curve) incident radiation on a cold medium. The full circle denotes $T(\tau_0=1,\Omega=1)$ calculated by the two-flux model. (b) Reflection coefficients R_∞ for a cold infinitely thick medium calculated as a function of albedo Ω. Curves are identified as in (a) [Reiss 1981a]

method have been reported for instance by Jaenicke [1956], Weber [1957], Kortüm and Oelkrug [1964], and Cabannes et al. [1979].

A diagrammatic analog to Fig.3.3b can be calculated if scattering is anisotropic; see below and Sect.3.2. The validity of the two-flux model, also in the case of anisotropic scattering, has recently been investigated [Brewster and Tien 1982a] by comparison with exact results obtained from the discrete ordinate method (Sect.3.2). This investigation is of particular interest because all required parameters were calculated from Mie theory, i.e. no free parameters were applied. With spherical particles (n = 1.21, x = 1) a deviation of at the most 10% is reported for transmission and reflection coefficients if optical thickness τ_0 is between 0.1 and 100. The corresponding backscattering factor b equals 0.43, however, which indicates almost isotropic scattering. If x = 50 the deviations are considerably larger (-30% to -50% from the exact results) because b = 0.14 (i.e., forward scattering dominates).

Obviously the deviations increase if anisotropic scattering increases, which is not properly accounted for by the heuristic introduction of a backscattering factor in (3.6a,b). Using a delta phase function in the two-flux model expressions, Truelove [1984] showed that the above listed deviations are removed when performing the calculations for the same set of variables and parameters as used by Brewster and Tien [1982a]. This finding has naturally led to a considerable reassessment of the two-flux model. Most of the work using the two-flux model is restricted to cold media (no emission term) and to the calculation of transmission, reflection or backscattering coefficients.

Few investigations have been presented that take into account a thermally conducting medium [Chan and Tien 1974b; Wiesmann 1983]. The latter reference uses also spectrally dependent optical parameters.

3.2 Discrete Ordinates

The discrete ordinate method is a generalization of the two-flux model: if directional intensity i' is strictly isotropic, it is sufficient to analyze the radiation field only in one direction (with two orientations). If, however, an anisotropic directional intensity results from an at least partially transparent medium and from anisotropic scattering, analysis of a radiation field in more than one direction is necessary. As a consequence we have instead of the single equation (2.44) a set of 2n transfer equations (index Λ again omitted)

$$\mu_i \cdot \frac{di'(\tau,\mu_i)}{d\tau} = - i'(\tau,\mu_i) + (1-\Omega)\cdot i_b' + (\Omega/2)\cdot\int_{-1}^{+1} i'(\tau,\mu_i)\cdot\Phi(\mu_i,\mu)\cdot d\mu \tag{3.14}$$

for the directions μ_i ($i=\pm1,\pm2,\ldots\pm n$), $\mu_{-i} = -\mu_i$. A solution of the system of equations (3.14) can be found if the scattering integral is replaced by the expansion

$$\int_{-1}^{+1} i'(\tau,\mu_i)\cdot\Phi(\mu_i,\mu)\cdot d\mu = \sum_{j=1}^{n} a_j\cdot f(\tau,\mu_j) \ . \tag{3.15}$$

In this equation the a_j denote weights that can be obtained from a suitable quadrature formula and an optimum division of the quadrature interval [-1,1]. Figure 3.4 illustrates the (optimum) directions μ_i that result from application of Gauss' quadrature formula in the case of three- and five-flux calculations. The scattering phase function of a glass fibre is given by the solid curve. Note that the choice of the μ_i does *not* depend on the phase function but follows exclusively from the zeros of the Legendre polynominals. The method is treated in detail by Chandrasekhar [1960 pp.54-69] and Kourganoff [1952 pp.101,102].

Kaganer [1969b] demonstrates the superiority of this method over the two-flux model when calculating three- and five-flux solutions for transmission and reflection coefficients using a Markov integration formula (the odd number of divisions μ_i is due to inclusion of μ = 0; the corresponding differential equation of the system (3.14)

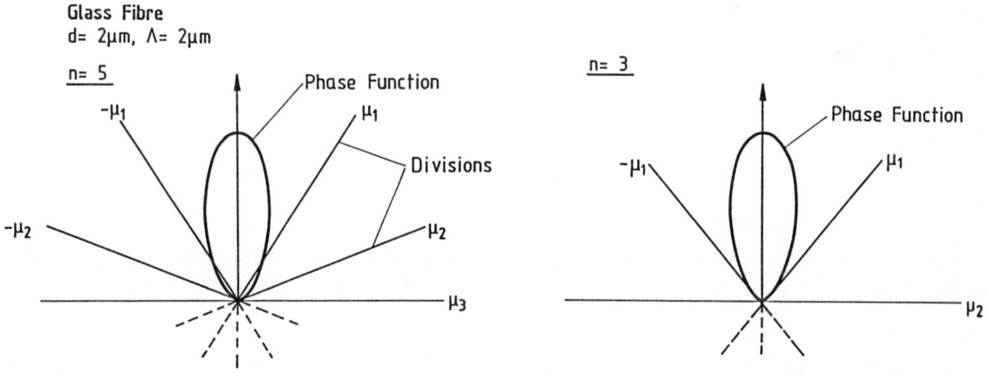

Fig.3.4. Single scattering phase function Φ for strong anisotropic scattering ($\bar{\mu}$=0.84) of unpolarized light (Λ=2μm), and divisions (directions) μ_i used for three- and five-flux approximations. After [Caps et al. 1984]. Note that anisotropy of intensity and anisotropy of scattering phase function are not identical in a multiple scattering process

transforms into an algebraic equation, which improves the accuracy of the method but adds no additional difficulties). However, Kaganer's analysis is confined to isotropic scattering. Three-flux solutions for cold media transmission coefficients and reflectivities R_∞ have already been given in Figs.3.3a,b for comparison with the corresponding two-flux model solutions. For $\Omega = 1$ and $\tau_0 = 1$ we have $T = 0.5714$. This is in considerably better agreement with the exact value than the two-flux result ($T = 1/2$, Sect.3.1). Extinction, absorption and scattering coefficients E, A and S are thus more precisely determined from optical experiments if a three- or five-flux solution is used instead of the two-flux model.

Figure 3.5 shows a comparison between Monte Carlo simulations and five-flux model solutions for transmission coefficients in cold media. The latter are obtained using again a Markov integration formula for transmission coefficients as a function of optical thickness and for different anisotropy factors μ [Caps et al. 1984]. The good agreement shows that this five-flux model solution is very suitable

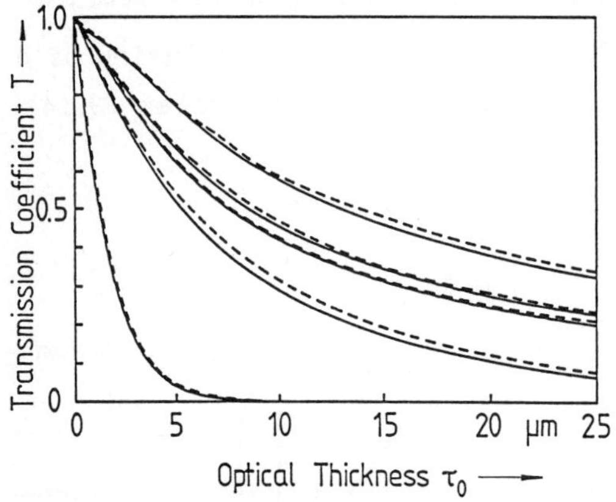

Fig.3.5. Comparison between five-flux approximation (dashed curves) and Monte Carlo simulation (solid curves) of transmission coefficients (diffuse incidence) calculated for a cold anisotropically scattering fibrous medium (particle diameter 2 μm) at wavelengths Λ = 2.5, 3, 4, 5 and 6 μm (counted from top to bottom) with corresponding anisotropy coefficients $\Omega \cdot \mu$ = 0.86, 0.79, 0.78, 0.72 and 0.26, respectively, as a function of optical thickness τ_0 [Caps et al. 1984]. Note that $T(\tau_0)$ follows approximately the two-flux model predictions $T(\tau_0) = 1/(1+\tau_0)$ at $\Lambda \leq 5$ μm, where scattering is strong, and $T(\tau_0) = \exp(-2\tau_0)$ at $\Lambda = 6$ μm (bottom curve), where absorption dominates (Figs.4.14,17). Reprinted by permission of Pergamon Journals Ltd.

for comparison of calculated transmission coefficients with experimental results, also in the case of strong anisotropic scattering. Comparisons of five-flux model calculations for reflectance and transmittance with experiments are given in Figs.4.14,15. High order discrete ordinate calculations are also reported by Brewster [1981], Roux and Smith [1981], and Maheu et al. [1984] (Brewster's thesis contains a computer program for this purpose).

In the work reviewed in Sect.3.3.3 a scaling factor (1-g) applied to the extinction coefficient or to the optical thickness is calculated. This factor has frequently been used in the literature to take into account anisotropic scattering. According to the scaling of τ_0, $\dot{q}_{Rad}$ in an anisotropically scattering medium of optical thickness τ_0 equals the $\dot{q}_{Rad}$ in an isotropically scattering medium if the optical thickness of this medium is $\tau_0 \cdot (1-g)$. The question then arises whether it is possible, by analogy, to use the well-known n-flux expressions calculated for isotropic scattering [Kaganer 1969b] with scaled τ_0 and albedo Ω, in order to account for anisotropic scattering in a simplified though accurate manner. Caps [1985] and also Baumeister [1986] have proven, by comparison with Monte-Carlo simulations, that this is indeed the case: very good agreement is reported for calculated (scaled) transmission and reflection coefficients T, R and R_∞ of cold media even if g takes a value up to 0.9. Clearly, these results have enormously simplified the calculation of T, R and R_∞, which are frequently measured quantities.

3.3 Formal Solution, its Numerical Calculation and its Consequences for Nontransparent Media

3.3.1 Derivation of Radiative Flow and Temperature Profile

We will now assume a grey, isotropically scattering[4] medium that is contained between two plane isothermal walls extending to infinity (Fig.2.13). The optical thickness $\tau(s)$ is given by $\tau(s) = \tau(x)/\cos\beta =$

[4] We could immediately consider a linear anisotropically scattering grey medium by inclusion of a scattering integral that is weighted by an anisotropy factor [Dayan and Tien 1975; Yuen and Wong 1980]; this will be dealt with in Sect.3.3.3.

$\tau(x)/\mu$. As a consequence, a formal solution to (1.2) or (2.44) reads

$$i'(\tau,\beta) = i'(0)\cdot\exp(-\tau/\mu) + \int_0^{\tau} I'(\tau^*)\cdot\exp[-(\tau-\tau^*)/\mu]\cdot\frac{d\tau^*}{\mu} \qquad (3.16)$$

where the source function $I'(\tau)$ collects the emission term $(1-\Omega)\cdot i_b'$ and the scattering integral in (2.44).

Since optical thickness τ_0 of nontransparent media is necessarily large, only solutions for $\tau_0 \to \infty$ of (3.16) are relevant for our discussion. However, a few important results also valid for smaller τ_0 will be reviewed.

If we assume $\Omega=0$, radiative equilibrium calls for

$$\sigma T^4(\tau) = (1/4)\int_0^{4\pi} i'(\tau,\omega)\cdot d\omega\ . \qquad (3.17)$$

The integral can be split into two parts by dividing the integration interval

$$\sigma T^4(\tau) = (1/4)\int_{\frown} i_+'(\tau,\omega)\cdot d\omega + (1/4)\int_{\smile} i_-'(\tau,\omega)\cdot d\omega\ . \qquad (3.18)$$

Note that this step is not equivalent to the derivation of the two-flux model because here $i'(\tau)$ is not assumed isotropic. For both components i_+' and i_-' we have formal solutions that follow immediately from (3.16).

If the thermal emissivity of the boundaries is that of a black body, we have

$$i_+'(0) = \frac{\sigma\cdot T_1^4}{\pi} \qquad (3.19a)$$

$$i_-'(0) = \frac{\sigma\cdot T_2^4}{\pi}\ . \qquad (3.19b)$$

Note that (3.19a,b) do not specify the temperatures $T(x')$ and $T(y')$ of the outermost layers of the medium but the directional radiation emitted from the walls, as a boundary condition. A temperature slip $[T_1-T(x')]$ or $[T(y')-T_2]$ may thus occur, disappearing only if the medium is either conducting or if the optical thickness τ_0 goes to infinity. If the medium is conducting, a gradient dT/dx must exist everywhere in its interior. If τ_0 is very large, on the other hand, the medium acts as a heat mirror. As a consequence, the temperatures $T(x')$ and $T(y')$ must approach very closely the wall temperatures T_1 and T_2, respectively. However, if the medium is nonconducting and if τ_0 is small, a temperature slip is mandatory. Otherwise, no radiation could be coupled into the medium because $q_{Rad}^{Wall} = \sigma \cdot \epsilon_{eff} \cdot [T_1^4 - T^4(x')]$ $\rightarrow 0$ if $T_1 \rightarrow T(x')$, where ϵ_{Eff} is an effective thermal emission coefficient.

Inserting these boundary conditions in the expressions for i_+' and i_-' (3.18) can be rewritten as follows

$$T^4(\tau) = (1/2) \int_0^{+1} \left\{ T_1^4 \cdot \exp(-\tau/\mu) + \int_0^{\tau} T^4(\tau^*) \cdot \exp[-(\tau-\tau^*)/\mu] \cdot \frac{d\tau^*}{\mu} \right.$$

$$\left. + T_2^4 \cdot \exp[-(\tau_0-\tau)/\mu] + \int_{\tau}^{\tau_0} T^4(\tau^*) \cdot \exp[-(\tau^*-\tau)/\mu] \cdot \frac{d\tau^*}{\mu} \right\} \cdot d\mu \,. \qquad (3.20)$$

A considerable simplification results if we apply the well-known exponential integral functions [Chandrasekhar 1960 pp.373-374]

$$E_n(x) = \int_0^1 \mu^{n-2} \cdot \exp(-x/\mu) \cdot d\mu \qquad (3.21)$$

to rewrite the exponentials in (3.20)

$$T^4(\tau) = (1/2) \left[T_1^4 \cdot E_2(\tau) + \int_0^{\tau} T^4(\tau^*) \cdot E_1(\tau-\tau^*) \cdot d\tau^* \right.$$

$$\left. + T_2^4 \cdot E_2(\tau_0-\tau) + \int_{\tau}^{\tau_0} T^4(\tau^*) \cdot E_1(\tau^*-\tau) \cdot d\tau^* \right] . \qquad (3.22)$$

In dimensionless form, the temperature distribution is given as a Fredholm integral equation of the second kind [Heaslett and Warming 1965]

$$\phi_b = \frac{T^4(\tau) - T_2}{T_1^4 - T_2^4} = \frac{E_2(\tau)}{2} + \int_0^{\tau_0} \phi_b(\tau^*)\cdot E_1(|\tau-\tau^*|)\cdot d\tau^* . \tag{3.23}$$

After numerical calculation of ϕ_b from (3.23) and because conservation of energy requires $\dot{q}_{Rad}(\tau)$ = const for all τ, i.e. also for τ=0, the dimensionless radiative flux density is given by[5]

$$\Psi_b = \frac{\dot{q}_{Rad}}{\sigma\cdot(T_1^4 - T_2^4)} = 1 - 2\int_0^{\tau_0} \phi_b(\tau^*)\cdot E_2(\tau^*)\cdot d\tau^* . \tag{3.24}$$

Figures 3.6,7 illustrate dimensionless temperature distribution and radiative flux ϕ_b and Ψ_b for different optical thicknesses τ_0 and black walls.

Equation (3.24) can be rearranged to give

$$\dot{q}_{Rad} = \Psi_b\cdot\sigma\cdot(T_1^4 - T_2^4) . \tag{3.25}$$

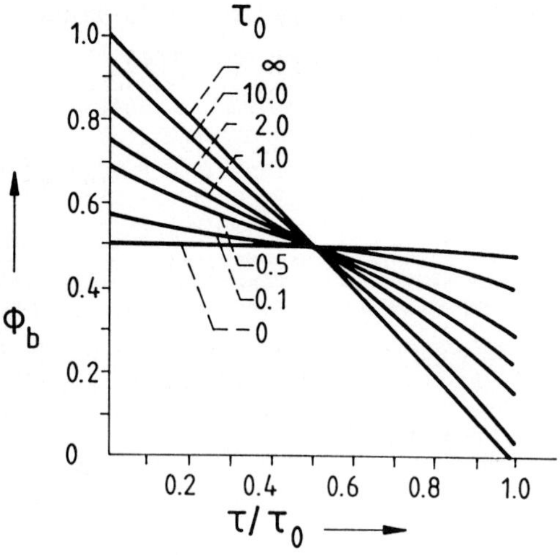

Fig.3.6. Dimensionless temperature profile ϕ_b in a grey absorbing/emitting gas contained between infinite black parallel plates, calculated for different optical thickness τ_0 from (3.23). From [Heaslett and Warming 1965] reprinted by permission of Pergamon Journals Ltd.

---

[5] The index of $E_3(\tau^*)$ given by Siegel and Howell [1972 p.452] is in error.

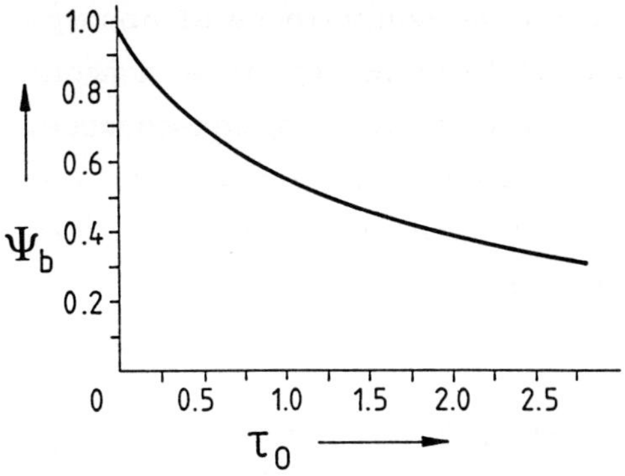

Fig.3.7. Dimensionless radiative flux density Ψ_b in a grey absorbing/emitting gas (same geometry as in Fig.3.6), calculated as a function of optical thickness τ_0 from (3.24). From [Heaslett and Warming 1965] reprinted by permission of Pergamon Journals Ltd.

Since $\Psi_b \rightarrow (4/3)/(1.42089 + \tau_0)$ if $\tau_0 >> 1$, we have the most important result

$$\dot{q}_{Rad} = \frac{4\sigma}{3} \cdot \frac{T_1^4 - T_2^4}{\tau_0} \quad \text{if} \quad \tau_0 >> 1 \; . \tag{3.26}$$

Since the $E_n(\tau) \rightarrow 0$ if $\tau \rightarrow \infty$, all terms in the previous equations that are products of constants and the $E_n(\tau)$ can be interpreted as "local" quantities. Conversely, if radiative transfer is described by a propagation law that used only local quantities or their derivatives, this law will lead us to (3.26) (Sect.3.4).

Extensions of these results to arbitrary albedo Ω introduces no additional difficulties. The source function

$$I'(\tau) = (1-\Omega)\cdot i_b'(\tau) + \frac{\Omega}{4\pi}\cdot \int i'(\tau,\beta)\cdot d\omega = (1-\Omega)\cdot \frac{\sigma\cdot T^4(\tau)}{\pi} + \Omega\cdot i'(\tau) \tag{3.27}$$

reduces to

$$I'(\tau) = i_b'(\tau) \tag{3.28}$$

because of radiative equilibrium [$i'(\tau)$ is a spatial average of $i'(\tau,\beta)$]

$$i'(\tau) = \frac{\sigma\cdot T^4(\tau)}{\pi} \; . \tag{3.29}$$

In these equations, $\pi\cdot i'(\tau)$ denotes the total, i.e. emitted and scattered, energy that is incident (from all directions) on a volume element. Equation (3.28) explicitly states the equivalence of absorption/emission and isotropic scattering for extinction in a nonconducting medium because $I'(\tau)$ does not depend on Ω. As a consequence, also temperature distribution and radiative flux density do not depend on albedo. All previous expressions that have been obtained for $\Omega = 0$ remain valid if we simply replace

$$\tau = \int_{s_1}^{s_2} A(s^*)\cdot ds^* \quad \text{by} \quad \tau = \int_{s_1}^{s_2} E(s^*)\cdot ds^* \ .$$

3.3.2 Viskanta's Solutions for a Grey Medium, for Isotropic Scattering and Temperature-Independent Parameters

The situation changes considerably if the medium is conducting, too. Inspection of (1.15) immediately shows that the factor $(1-\Omega)$ does not cancel. As a consequence, the temperature distribution $\Theta(\tau)$ calculated from (1.15) depends on Ω. Since i_{Ab}' in (2.44) depends on temperature, $\dot{q}_{Rad}$ is a function of Ω (it is rather a "natural consequence" that radiative heat flow depends on the "optical" parameter Ω). Since in addition conductive heat flow depends on $d\Theta/d\tau$, we have the surprising result (at first sight) that a *calorimetric* quantity also depends on an *optical* parameter. An analogy that is the reverse of this finding is immediately obvious: as a result of (1.15), $\Theta(\tau)$ depends also on N_1, which is a function of thermal conductivity. Therefore, the *optical* quantity "radiative heat flow" depends on a *calorimetric* parameter. Of course, these dependencies are merely the consequence of conservation of energy. It is seen that in a conducting medium, absorption/emission and isotropic scattering are no longer equivalent ("degenerate") with regard to extinction.

The complete expressions for $\eta(\tau)$, $\dot{q}_{Rad}(\tau)$ and $d^2\Theta(\tau)/d\tau^2$ are given by Viskanta [1965], and Sparrow and Cess [1966 pp.250-253] and will not be repeated here. While it was sufficient to prescribe *intensities* as boundary conditions in the nonconducting medium, now the boundary conditions must specify *temperatures* because $d\Theta/d\tau$ exists everywhere in the medium including $\tau = 0$ and $\tau = \tau_0$: at these locations, $\Theta = 1$ or $\Theta = \Theta_2$, respectively. There are *no* temperature jumps in a conducting medium.

In a series of publications, Viskanta and Grosh studied systematically the influence of the parametric functions $\epsilon_{1,2}$, τ, τ_0, N_1 and Ω on temperature profile and radiative and conductive heat flow in grey absorbing/emitting, isotropically scattering, heat conducting media [Viskanta and Grosh 1962a,b,1964; Viskanta 1965]. At that stage, their papers opened a new (predominantly numerical) discipline in investigating radiative transfer. Some of the most remarkable results are collected in Fig.3.8a-f [note that the definition of the parameter $N=4n^2 \cdot \sigma \cdot T^4/(\lambda_{Cond} \cdot E \cdot T_1)$ given by Viskanta [1965] is different from the conduction/radiation parameter N_1 used in (1.15)].

Figure 3.8a illustrates that in a conducting medium, the dimensionless radiative flux density ϕ is not necessarily a constant (by conservation of energy, only the total heat flow $\dot{q}$ is constant at all τ). Figure 3.8b reveals that even if ϕ should vanish at $\tau/\tau_0 = 0$ and $\tau/\tau_0 = 1$, because the wall thermal emissivities $\epsilon_{1,2} = 0$, the interaction of radiation with conduction causes a nonvanishing ϕ for all $0 < \tau/\tau_0 < 1$. The dependence of ϕ and of the dimensionless temperature Θ on albedo Ω is illustrated by the examples given in Fig.3.8c,d. If $\Omega = 1$, the radiation field decouples completely from the temperature field so that ϕ is again constant and Θ is linear in τ/τ_0. Fig.3.8e shows how the dimensionless total heat flux Ψ depends on the albedo Ω. This result explicitly shows that the "degeneracy" of absorption/emission and isotropic scattering no longer exists if the medium has a nonvanishing thermal conductivity. Fig.3.8f illustrates that wall emissivities $\epsilon_{1,2}$ may exert an influence on Θ if the optical thickness of the medium is low and if radiation clearly dominates over conduction.

Viscanta's (iterative) methods[6] do not comprise a study of the influence of temperature dependent parameter functions $\lambda_{Cond}(T)$, $E(T)$,

[6] Mihalas [1967 pp.1-52] deals with numerical methods to calculate temperature distributions in absorbing/emitting and scattering media.

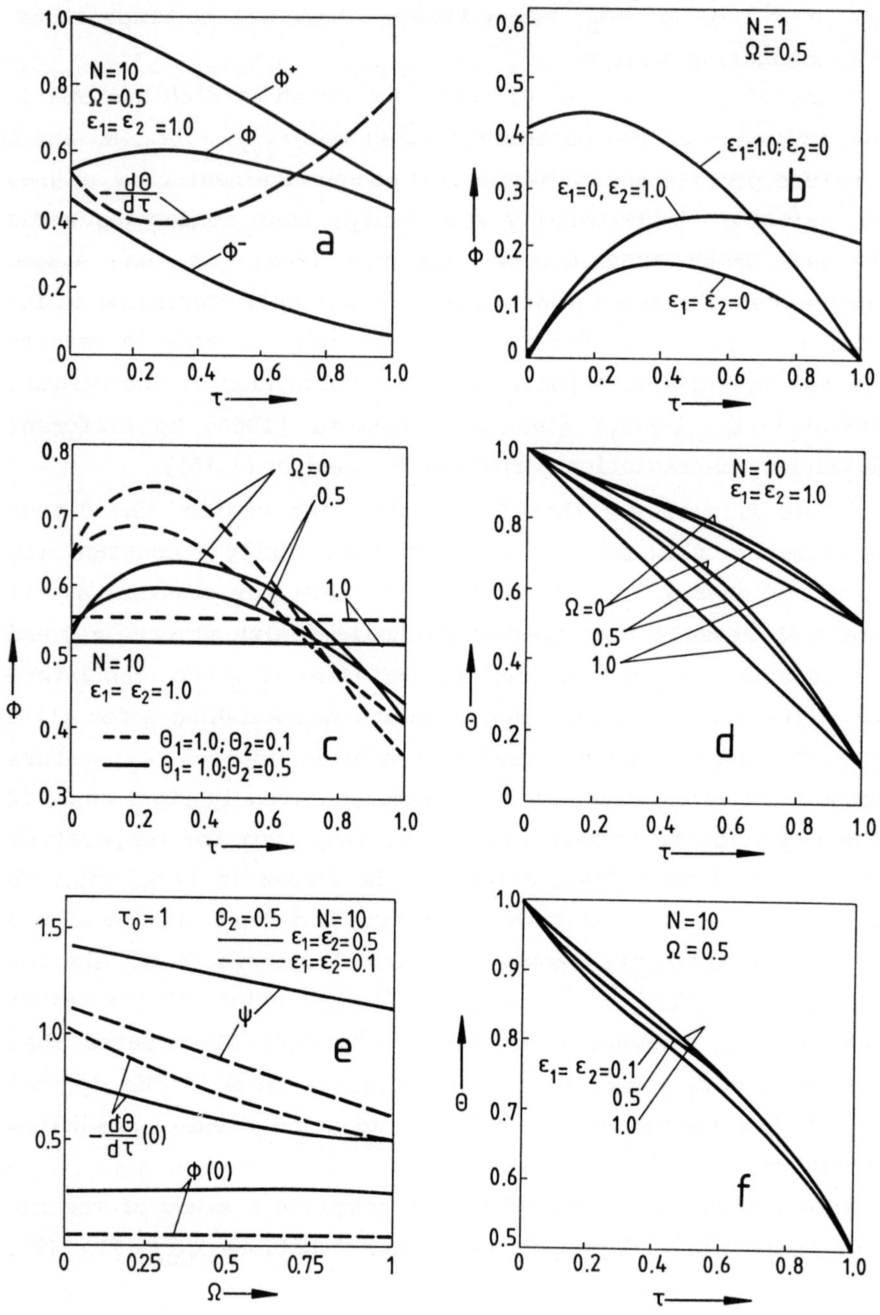

Fig.3.8a–f. Caption see opposite page

Fig.3.8. Viskanta's results for radiative transfer in a homogeneous, grey, absorbing/emitting, isotropically scattering, conductive, planar medium. Calculated curves are given as a function of local optical thickness τ and for different $N = 4n^2 \cdot \sigma \cdot T^4/(\lambda_{Cond} \cdot E \cdot T_1)$, albedo Ω, thermal emissivities $\epsilon_{1,2}$ of side walls 1 and 2, and dimensionless temperatures $\Theta_{1,2} = T_{1,2}/T_1$. All diagrams from [Viskanta 1965]; for comparison with other work: $N = 1/N_1$ according to (1.16).

(a) Dimensionless local total radiative heat flow density $\phi = \dot{q}_{Rad}/(n^2 \cdot \sigma \cdot T^4)$ and dimensionless gradient $d\Theta/d\tau$ (dimensionless conductive flow density) calculated for $\tau_0 = 1$, $\Theta_1 = 1$, $\Theta_2 = 0.5$. Note that conservation of energy only requires $\dot{q}_{Rad} + \dot{q}_{Cond} = const$, i.e. a local variation of both flux densities may occur.

(b) Dependence of ϕ on wall emissivities (τ_0=1). Note that ϕ does not vanish for $0<\tau<1$ even if $\epsilon_1 = \epsilon_2 = 0$ because of coupling of radiative and conductive flow.

(c) Dependence of ϕ on albedo (τ_0=1). Ω = 0 and 1 correspond to strongest and weakest interaction between radiation and conduction, respectively.

(d) Dependence of dimensionless temperature profiles $\Theta = T/T_1$ on albedo (τ_0=1, Θ_1=1). As in (c), Θ reflects for Ω = 0 or 1 the strongest or weakest interaction of radiation with conduction, respectively. The profile depends exclusively on conduction properties if Ω = 1. The two sets of curves refer to Θ_2 = 0.1 or 0.5, respectively.

(e) Dependence of dimensionless total heat flux density $\Psi = \dot{q}/(\lambda_{Cond} \cdot E \cdot T_1)$, $\phi(0)$ and $d\Theta/d\tau(0)$ on albedo (τ_0=1, Θ_1=1, Θ_2=0.5, $\epsilon_1=\epsilon_2$=1). Note the decrease of total $\dot{q}$ with increasing Ω, i.e. absorption and scattering are *not* equivalent mechanisms for total extinction if the medium has nonvanishing conduction properties.

(f) Dependence of Θ on wall emissivities (τ_0=1). If ϵ is small, radiation emitted from boundaries is reduced whereas $d\Theta/d\tau$ near the walls must increase by conservation of energy

$\Omega(T)$ and $\epsilon_{1,2}(T)$. The same applies to recent work of other authors[7] who preferentially study the influence of anisotropic scattering on radiative and conductive flux and temperature profile (see below). As we have seen in Chap.2, neglect of temperature dependence, in particular of λ_{Cond}, is not a suitable description of physical reality. In Sects.3.4.4 and 5.2 temperature dependent parameters will be used

[7] Compared with the 'formal' parameter tests performed in Viskanta's work, Tong and Tien [1980,1983], and Wang and Tien [1983] applied parametric functions that are calculated from Mie's theory. However, they use again the two-flux model with inclusion of anisotropic scattering, which is taken into account by a linear anisotropic phase function. Maheu et al. [1984] also used Mie's theory for determination of parameter functions when calculating transmission and reflexion properties from a four-flux model in a cold medium, however, without accounting for conduction.

when calculating radiative and conductive heat flow and temperature profiles by diffusion methods.

3.3.3 Inclusion of Anisotropic Scattering by LAS Models; Importance of Wall Emissivity

Another important step is the inclusion of anisotropic scattering in solutions of the transfer equation by means of linear anisotropic scattering (LAS) models [Yuen and Wong 1980]. Investigations into the influence of LAS on radiative heat flow and temperature profiles yielding similar results have already been presented by Dayan and Tien [1975] and later by Azad and Modest [1981]. However, these authors do not consider thermally conducting media.

Although the mathematical methods to solve the transfer and energy equations are different from Viskanta's procedure, clearly a closer relation between the work of Yuen and Wong [1980] and Viskanta's work exists, from a physical point of view, because both consider heat conducting media (that is, progress of Yuen's work in relation to Viskanta's calculations is the inclusion of anisotropic scattering). Yuen uses a power series expansion of Θ and Θ^4 that allows subsequent analytical integration of the terms that appear in transfer and energy equation with the help of the $E_n(\tau)$. Anisotropic scattering is included by the introduction of another scattering integral $(\Omega/2)\cdot a\cdot\mu\cdot\int i'(\tau,\mu)\cdot\mu\cdot d\mu$ that is weighted by an anisotropy factor a. This additional contribution to the source function arises from an expansion of the scattering phase function $\Phi(\mu)$ as done in (2.65). Using only the first order term,

$$\Phi(\mu) = 1 + a\cdot\mu \tag{3.30}$$

we introduce this expansion in the usual scattering integral$(\Omega/2)\cdot\int i'(\tau,\mu)\cdot\Phi(\mu)\cdot d\mu$ [Eq.(3.30) is known as the LAS phase function, hence the name "LAS model" for this approximation]. Comparing this with (2.66) we see that $a = \omega_1/\Omega = 3\bar{\mu}$.

Figure 3.9 shows Yuen's dimensionless temperature profile [apart from the constant n^2, the definition of the conduction/radiation par-

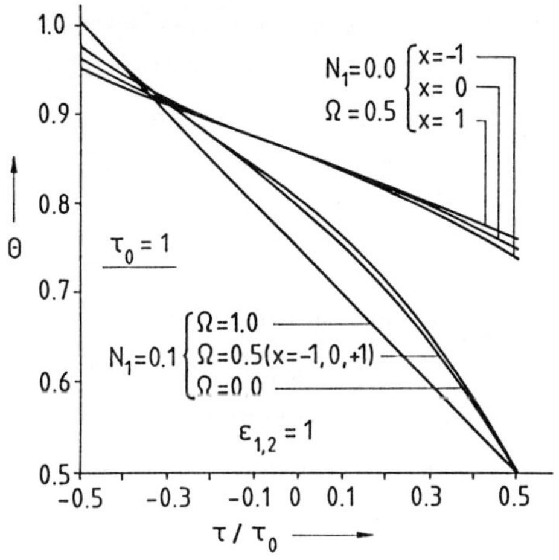

Fig.3.9. Effect of anisotropic scattering on dimensionless temperature profile calculated for different conduction/radiation parameters $N_1 = \lambda_{Cond} \cdot E/(4\sigma \cdot n^2 \cdot T_1{}^3)$ using $n^2 = 1$, albedo Ω and anisotropy parameter $x = \omega_1/\Omega = 3\,\mu$ [Yuen and Wong 1980]

ameter $N_1 = \lambda_{Cond} \cdot E/(4\sigma \cdot T_1{}^3)$ is identical to (1.16)]. The dependence of Θ on $x = a$ is rather weak compared to the effect of the albedo Ω.

While Viskanta [1965] predicts a decrease of total heat flow $\dot{q}$ that generally occurs if Ω increases (Fig.3.8e), Yuen's improved analysis shows that this is strictly true only if scattering is isotropic ($a = \omega_1/\Omega = 0$). If for instance Ω=1, $\dot{q}$ can be larger than for $\Omega = 0$ if a equals 1 (forward scattering). If the wall emissivities ϵ are large, the influence of Ω is small. For small ϵ and $0.01 \leq N_1 \leq 1$, $\tau_0 = 1$ or 10, however, $\dot{q}$ can be smaller if Ω=1, a=1 than for Ω=0. Furthermore, Yuen shows that the additive approximation can be seriously in error if $\epsilon = 0.1$ and $\tau_0 = 1$.

The importance of small wall emissivities, with regard to the amount of net radiative flow, can easily be understood: small ϵ means large reflectivity, which must exert an influence on radiative transfer analogous to that of forward scattering if the optical thickness is small. If the optical thickness increases and ϵ is large, only the influence of the anisotropy factor a remains.

Obviously, the influence of wall emissivities (or reflectivities) is considerable in transparent media. A formal study of parameter dependencies should thus also include a test of how diffuse or specular reflection enter the results for $\dot{q}$, $\dot{q}_{Rad}$ and T(x): radiation that is specularly reflected at small angles β (Fig.2.13) travels on average a shorter path between two reflections than in the diffuse reflection case. Accordingly, this radiation suffers less extinction, and the number of reflections increases before this radiation is completely extinguished. Contrary to these expectations, Yuen and Wong [1981] find only minor differences in $\dot{q}_{Rrad}$ and T(x) between the two reflection regimes. No dependence on anisotropy factor a is observed.

The influence of wall emissivity on $\dot{q}_{Rad}$ has also been investigated by designing an "effective emission coefficient" ϵ^*: Scheuerpflug et al. [1985] introduced the semiempirical expression $\epsilon^* = 1 - (1-\epsilon)\cdot\exp\{-N_1/(2N_1+0.04)\cdot\arctan[\tau_0\cdot(1+0.02/N_1)]\}$ instead of the familiar emissivity ϵ into the grey-wall complement of (3.25), i.e. (1.12)

$$\dot{q}_{Rad} = \frac{\sigma\cdot(T_1^4 - T_2^4)}{1/\epsilon_1 + 1/\epsilon_2 - 1 + 3\tau_0/4}$$

in order to calculate $\dot{q}_{Rad}$ in transparent SiO_2 aerogel and for small $\epsilon_{1,2}$. That is, they used for the aerogel

$$\dot{q}_{Rad} = \frac{\sigma\cdot(T_1^4 - T_2^4)}{1/\epsilon_1{}^* + 1/\epsilon_2{}^* - 1 + 3\tau_0/4}$$

instead of (1.12). The aim of this approximation is to account for the coupling of conduction and radiation without explicitly solving (2.44) and (1.15) and to use the additive approximation. The ϵ^* simulate the existence of a "bright" boundary layer in front of the rather dark wall (of low emissivity) that radiates an increased amount of heat into the interior of the medium (because of the presence of conduction). Good agreement is found between calculations and experiment assuming a grey medium. If, in addition, τ_0 is replaced by its scaled value $\tau_0{}^* = \tau_0\cdot(1-g)$, agreement is found also with Yuen's results for an anisotropically scattering, conductive medium [Caps 1985]. Of course, the full value of these approximations becomes obvious in transparent, not in nontransparent media.

Another interesting method to solve the radiative transfer problem (without the inclusion of conduction) uses the derivatives $d^n/d\tau^n$ of (2.44). This method establishes a differential equation, a successive approximation solution of which leads in first order to [Yuen and Tien 1980]

$$\dot{q}_{Rad} = \frac{\sigma \cdot (T_1^4 - T_2^4)}{1 + (3 - \omega_0 \cdot x_0) \cdot \tau_0 / 4} \quad (3.31)$$

where $x_0 = a$. This result is of particular importance because it has led us directly to a scaled, i.e. an "effective" value $(3/4-\omega_0 \cdot x_0/4) \cdot \tau_0$ of τ_0. Eq.(3.31) is another example among a group of LAS approximations. In this equation $\omega_0 \cdot x_0/4 = \omega_1/4 = (3/4)\Omega \cdot \mu$ so that the factor in front of τ_0 equals $3(1-\Omega \cdot \mu)/4$. If τ_0 is large, we have

$$\dot{q}_{Rad} = \frac{4\sigma \cdot (T_1^4 - T_2^4)}{3(1-\Omega \cdot \mu) \cdot \tau_0} . \quad (3.32)$$

We will see in the next section that the same expression results also from the diffusion model.

Yuen's method rapidly converges to the exact solutions of the transfer problem. This is illustrated in Fig.3.10 for the dimensionless temperature.

Another test of the validity of the LAS approximation has been performed in Fig.3.11 [Caps et al. 1984] using transmission coefficients that are based on (3.31). Comparison is made between a Monte Carlo simulation and the LAS model for different phase functions. Obviously, the LAS model yields correct predictions also in the case of small τ_0 even when the scattering is strongly anisotropic.

3.3.4 Transient Temperature Profiles from Finite Element Calculations

Further numerical solutions include the finite element methods in absorbing/emitting, isotropically scattering and heat conducting media [Reddy and Murty 1978; Fernandes et al. 1980; Wu et al. 1980;

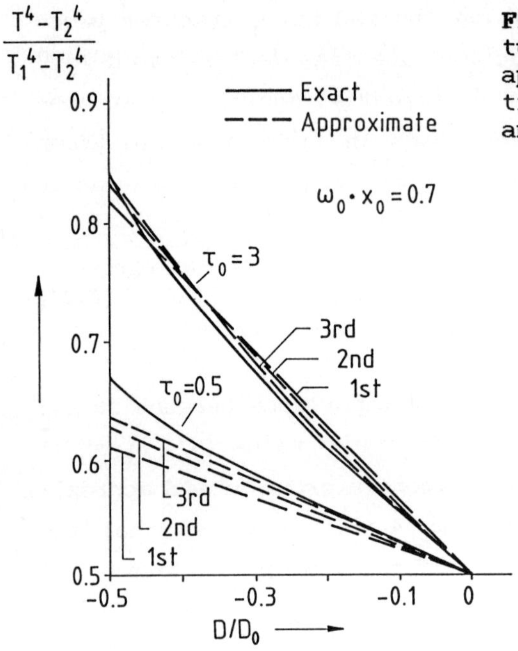

Fig.3.10. Comparison of temperature profiles from exact and approximate solutions with anisotropic scattering; see text [Yuen and Tien 1980]

Fernandes and Francis 1982]. These investigations involve different geometries and in particular transient heat transfer problems. Transient temperature profiles (assuming a fixed time) deviate the more from stationary solutions the larger the albedo Ω (Fig.3.12) [Fernandes et al. 1980]). This result can be interpreted as follows [Klemens 1985]: Scattered radiation travels in a diffusion-like manner but with the velocity of light to the heat sink. This radia-

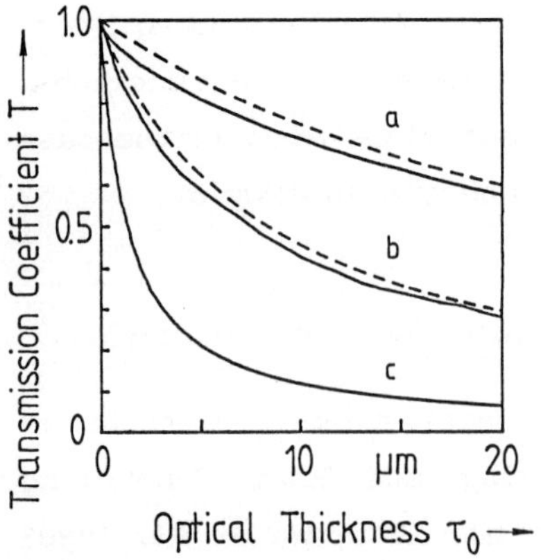

Fig.3.11. Comparison between LAS approximation (dashed curves) and Monte Carlo simulation (solid curves) of transmission coefficients (diffuse incidence) calculated for a cold strongly anisotropically scattering fibrous medium (curves a and b) and for isotropic scattering (curve c) as a function of optical thickness τ_0; curves a, b and c correspond to $\bar{\mu}$ = 0.955, 0.84 and 0, respectively. [Caps et al. 1984] reprinted by permission of Pergamon Journals Ltd.

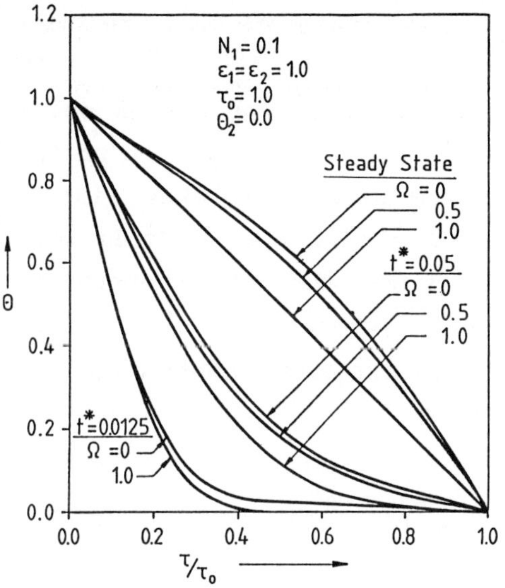

Fig.3.12. Dimensionless transient temperature profiles in a homogeneous, planar, absorbing/emitting, isotropically scattering, conductive medium calculated by finite element method [Fernandes et al. 1980] as a function of optical depth τ, albedo Ω and dimensionless time $t^* = \lambda_{Cond} \cdot E^2 \cdot t/(\rho \cdot C_p)$, t = time. Copyright AIAA (1980), reprinted with permission

tion is absorbed at the cold boundary but no thermal equilibrium has been established within the medium. As a consequence, there is a large temperature gradient near the hot boundary. This gradient disappears only gradually because "true" diffusion (absorption/emission) processes are considerably slower than elastic photon scattering (Sect.2.4.5). Nonstationary measurements of thermal conductivity can thus be greatly in error if the samples have strong (especially forward) scattering properties because heat flow and temperature distribution are only partly correlated in these cases (the correlation disappears completely if $\Omega = 1$).

3.3.5 Comments on "Linearization of the Temperature Profile" in Coupled Radiation-Conduction Problems

Note that almost all investigations mentioned in this and in the preceding subsections deal with grey media and grey boundaries. All emission terms in the transfer and energy equation thus simplify to the Stefan-Boltzmann law. Calculations that use spectral properties are very scarce: Wang and Tien [1983], Wiesmann [1983] who applies the two-flux model; and some papers on semitransparent media by Anderson and Viskanta [1973], Chupp and Viskanta [1974], and Domoto and Wang [1974]. All investigations except for Wiesmann's and those reported in Sect.3.4.4 and Chap.5 have used parameters that are independent of temperature.

Intermediate approaches between two-flux model or formal solutions and diffusion methods that result from a linearization of the temperature profile can be found in the work of Hamaker [1947], Walther et al. [1953], and Larkin and Churchill [1959]. It is well known, and follows immediately from (1.15), that the temperature profile is approximately linear if (besides possibly $\Omega = 1$) N_1 is very large. N_1 is very large if we have a large extinction coefficient *and* a large thermal conductivity. In this case

$$\lim_{N_1 \to \infty} \left(\frac{d^2\Theta}{d\tau^2}\right) = \lim_{N_1 \to \infty} \left\{\frac{1-\Omega}{N_1} \cdot \left[\Theta^4(\tau) - \frac{\eta(\tau)}{4}\right]\right\} = 0 \tag{3.33}$$

calls for $\Theta = C_1 + C_2 \cdot \tau$ (where $C_{1,2}$ are constants). Thus, a linearization of the temperature profile is equivalent to the assumption that an absorbing/emitting medium is nontransparent and that (temperature independent) conduction processes are large compared to radiative transfer. We will see in the next section that these premises lead us to the diffusion model of radiative transfer. However, although assumed in the analysis, linearity of the temprature profile was not experimentally verified in the work cited above. Larkin and Churchill [1959] investigate heat flow through glass fibres; Fig.3.13 [Reiss and Ziegenbein 1983] shows, however, that the profile is linear only if the medium is made nontransparent by heavy doping with Fe_3O_4 as an opacifier *and* if residual gas pressure is large.[8] In all other cases, curved temperature profiles result. The curvature is the stronger the smaller the fibre density, i.e. the smaller the extinction coefficient. It is thus very unlikely that linear temperature profiles can be observed in pure glass fibres or other pure substances.

Spectral extinction measurements (Sect.4.3) of pure glass fibres show in addition that scattering that would favour a linear temperature profile if λ_{Cond} were constant dominates only at wavelengths below 6 μm.

---

[8] I.e., it is experimentally verified that it is not sufficient to require large extinction coefficients *or* large thermal conductivity for establishing a linear temperature profile.

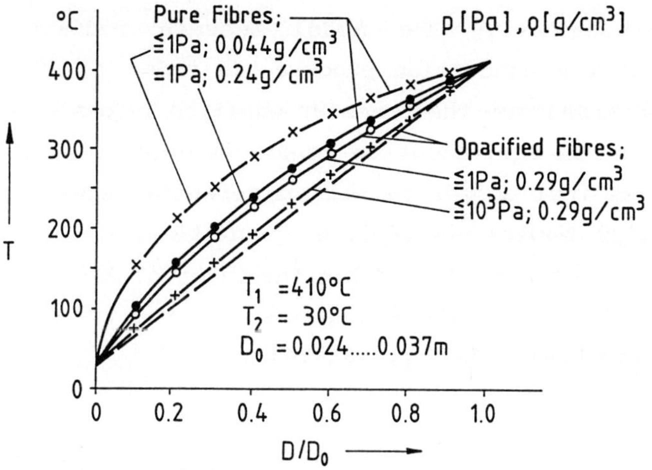

Fig.3.13. Experimental temperature profiles in pure and opacified, evacuated, homogeneous glass fibres; D_0 denotes total sample thickness [Reiss and Ziegenbein 1983]

3.4 The Diffusion Model

3.4.1 Derivation of Original Formulation

The diffusion model is another example of radiative transfer theories that were originally derived by astrophysicists. Like other approximate methods, the diffusion model reduces the integro-differential transfer equation to a simple differential transfer equation that can easily be integrated.

Let a homogeneous, purely absorbing medium with a very large extinction coefficient $E_\Lambda = A_\Lambda$ (at all wavelengths Λ) be given. The mean free path $\ell_{Rad,\Lambda}$ of photons equals $1/E_\Lambda$. If $\ell_{Rad,\Lambda}$ is small compared with the distance L over which significant temperature changes occur, a local intensity $i_\Lambda'(\tau_\Lambda)$ can originate only from a local temperature distribution. Radiation that is emitted from locations of different temperatures or is scattered from these locations will be greatly attenuated before it reaches location τ_Λ. A transport of radiation is thus possible only if the radiation travels by a series of "microsteps", i.e. absorption/emission (or scattering, see below) processes that occur only between immediate neighbours of the medium's constituents. Obviously, the model assumes radiative transfer to be essentially a conduction process. In order to show this, it is sufficient to transform the transfer equation into a pure differential equation that is equivalent to Fourier's empirical law (1.6 or 9). All quantities that appear in this differential equation must be local quantities or derivatives of local quantities.

Expansion of i_Λ' into a Taylor series within the interval $x-\ell_{Rad}/L \leq x \leq x+\ell_{Rad}/L$ and substitution of the derivatives into the transfer equation yields [Siegel and Howell 1972 p.469-474]

$$i_\Lambda' = i_{\Lambda b}' - \frac{\mu}{E_\Lambda} \cdot \frac{di_{\Lambda b}'}{dx} \tag{3.34}$$

if only first-order terms are considered. According to (3.34), the local, directional, spectral intensity i_Λ' depends only on the magnitude and gradient d/dx of the local, directional, spectral intensity $i_{\Lambda b}'$ of a black body. The directional intensity $i_{\Lambda b}'$ of a black body is isotropic. The extinction coefficient E_Λ is large at all wavelengths, per definition. The derivative $di_{\Lambda b}'/dx$ is small because local values of dT/dx are small, again per definition. As a consequence, the second term in (3.34) is negligible, i.e. $i_\Lambda'(\tau) \simeq i_{\Lambda b}'(\tau_\Lambda)$. Since $i_{\Lambda b}'(\tau_\Lambda)$ is isotropic, so must $i_\Lambda'(\tau_\Lambda)$. This conclusion is most important for nontransparent media.

Calculation of local radiative flux density $\dot{q}_{Rad,\Lambda}(\tau_\Lambda) = -\int i_\Lambda'(\tau) \cdot \mu \cdot d\mu$ yields Rosseland's differential equation [Rosseland 1931]

$$\dot{q}_{Rad,\Lambda} = - \frac{4}{3E_\Lambda} \cdot \frac{de_{\Lambda b}}{dx} \tag{3.35}$$

which couples local spectral flux density to the (local) gradient d/dx of hemispherical spectral emissive power of a black body. If the medium is grey, the formal similarity between Rosseland's law and (1.6 or 9) becomes more obvious

$$\dot{q}_{Rad} = \int_0^\infty \dot{q}_{Rad,\Lambda} \cdot d\Lambda = - \frac{4}{3E} \cdot \frac{d}{dx} \int_0^\infty e_{\Lambda b} \cdot d\Lambda = - \frac{4}{3E} \cdot \frac{de_b}{dx} , \tag{3.36a}$$

$$= - \frac{16\sigma\cdot n^2}{3E}\cdot T^3\cdot \frac{dT}{dx} \quad . \qquad (3.36b)$$

From this equation, a "radiative conductivity"

$$\lambda_{Rad} = \frac{16\sigma\cdot n^2}{3E}\cdot T^3 \qquad (3.37)$$

can be extracted. (This expression was also derived in a different way by van der Held [1952]). Note that the calorimetric quantity λ_{Rad} is only meaningful if the assumptions made above as to extinction properties at all wavelengths and temperature variation are justified.

Equation (3.36a) can be compared with the usual diffusion law $\dot{q} = D\cdot du/dx$ where u denotes radiative energy density, $u = (4n^3\cdot\sigma\cdot T^4)/c_0 = 4(n/c_0)\cdot e_b$, and D is the diffusion coefficient, $D = c_0/(3n\cdot E)$ [m^2/s]. The velocity of light in vacuum is denoted by c_0. Using $E=10^4$ m^{-1} and $n = 1$, we have $D = 10^4$ m^2/s. In a purely absorbing nonconducting medium, measurement of the diffusion depth $L = (D\cdot t)^{1/2}$ (t: time) could be used to determine an average speed by which the diffusion front proceeds. (In contrast to the comments made in Sec.2.4.5 [after (2.74a,b)], measurement of this speed would allow determination of the absorption coefficient if n were known).

If the medium is not grey, (3.35) can be integrated over $d\Lambda$ only within a finite inverval $\Delta\Lambda$:

$$\dot{q}_{Rad,\Delta\Lambda} = - (4/3) \int_{\Delta\Lambda} \frac{1}{E_\Lambda}\cdot\frac{\partial e_{\Lambda b}}{\partial x} \cdot d\Lambda \quad . \qquad (3.38)$$

If E_Λ is taken out of the integral, we have

$$\dot{q}_{Rad,\Delta\Lambda} = - \frac{4}{3E_{R,\Delta\Lambda}} \cdot \int_{\Delta\Lambda} \frac{\partial e_{\Lambda b}}{\partial x} \cdot d\Lambda \quad . \qquad (3.39)$$

In this equation, $E_{R,\Delta\Lambda}$ denotes a mean of the extinction coefficient E_Λ with respect to the interval $\Delta\Lambda$. The "Rosseland mean extinction coefficient" $E_{R,\Delta\Lambda}$ depends on temperature at least because $e_{\Lambda b}$ is a function of temperature

$$\frac{1}{E_{R,\Delta\Lambda}} = \frac{\int_{\Delta\Lambda} \frac{1}{E_\Lambda} \cdot \frac{\partial e_{\Lambda b}}{\partial x} \cdot d\Lambda}{\int_{\Delta\Lambda} \frac{\partial e_{\Lambda b}}{\partial x} \cdot d\Lambda} = \frac{\int_{\Delta\Lambda} \frac{1}{E_\Lambda} \cdot \frac{\partial e_{\Lambda b}}{\partial e_b} \cdot d\Lambda}{\int_{\Delta\Lambda} \frac{\partial e_{\Lambda b}}{\partial e_b} \cdot d\Lambda} = \frac{1}{E_R(T)} . \tag{3.40}$$

For a convenient formula to calculate the weight function $\partial e_{\Lambda b}/\partial e_b$ see Siegel and Howell [1972 p.474]. The weight function is illustrated in dimensionless form in [Caps et al. 1983a Fig.4].

Calculation of $E_R(T)$ is meaningful only if E_Λ is large at all wavelengths. It is not possible to correctly describe radiative transfer in semitransparent media by application of a mean extinction coefficient (see the discussion of Anderson and Viskanta [1973], who find deviations of up to 200%). Conversely, a large value of $E_R(\tau)$ does not necessarily imply large spectral values E_Λ at all wavelengths. The equivalence of absorption/emission and isotropic scattering for extinction in a nonconductive medium has been shown in Sect.3.3.1. As a consequence, it is allowed to use (3.36-40) also if $E_\Lambda = A_\Lambda$ is substituted by $E_\Lambda = A_\Lambda + S_\Lambda$ where S_Λ denotes the isotropic scattering coefficient. Note that this conclusion can be in error if the medium has conduction properties and if the different heat conduction modes are not uncoupled from each other.

3.4.2 Additive Approximation, Temperature Slip

From the end of Sect.3.4.1 it has to be concluded that the frequently used "additive approximation"

$$\dot{q} = \dot{q}_{Cond} + \dot{q}_{Rad} \tag{3.41}$$

not only assumes that the components $\dot{q}_{Cond}$ and $\dot{q}_{Rad}$ can be calculated independently of each other. In addition, it is assumed that absorption/emission and isotropic scattering are equivalent processes for extinction. The components $\dot{q}_{Cond}$ and $\dot{q}_{Rad}$ are independent of each other if they are not coupled by the temperature profile. The temperature profile does not couple heat flow components if it is linear (as in the pure scattering case). The conditions for linearity of the temperature profile have been illustrated in Sects.1.6 and 3.3.5: large extinction coefficient and large temperature independent thermal conductivity. Also the second condition is incorporated in the derivation of the diffusion model: if $\dot{q}$ = const, and because extinction is large, $-dT/dx$ can be small [i.e., nearly equal to $(T_1-T_2)/D$] everywhere in the medium only if λ is large. The premises for the application of a diffusion model *and* an additive approximation are thus the same.

Note that conditions necessary and sufficient for application of the diffusion model might not be sufficient in all cases for an application of the additive approximation. For example, let the temperature profile of type b in Fig.1.2 be given. Temperature profiles of this type (i.e., profiles with a small curvature) are typical for radiative diffusion processes. Let us assume that an experimentalist who wishes to calculate the total heat transfer $\dot{q}$ does not know this profile but only T_1, T_2, D and theoretical expressions for the conductivity components $\lambda_{Cond}(T)$ and $\lambda_{Rad}(T)$. If he applies the additive approximation to this situation, he calculates

$$\dot{q} = [\lambda_{Cond}(T_m) + \lambda_{Rad}(T_m)] \cdot \frac{T_1 - T_2}{D} ,$$

i.e. he necessarily assumes the temperature profile to be linear. Therefore, he has to calculate $\lambda_{Cond}(T)$ and $\lambda_{Rad}(T)$ at the mean temperature $T_m = (T_1+T_2)/2$ that exists at $x = D/2$. However, the really existing curvature of the temperature profile has to be taken into account. Because of conservation of energy, $\dot{q}(x) = \dot{q}$ is a constant, i.e. $\dot{q}$ is independent of the position $0 \leq x \leq D$. Therefore, $\dot{q}$ can be calculated for an arbitrary position x, e.g. for the coordinate x', where $-dT(x)/dx$ equals $(T_1-T_2)/D$. Since $T(x')$ is in general not equal to $T(D/2)$,

$$-\{\lambda_{Cond}[T(x')] + \lambda_{Rad}[T(x')]\} \cdot \frac{dT}{dx}(x') \neq \frac{[\lambda_{Cond}(T_m) + \lambda_{Rad}(T_m)] \cdot (T_1 - T_2)}{D}$$

by conservation of energy.

In terms of radiative conductivity, the total heat flux density $\dot{q}$ is given, in the diffusion approximation, by

$$\dot{q} = -\left[\lambda_{Cond}(T) + \frac{16\sigma \cdot n^2}{3E_R(T)} \cdot T^3\right] \cdot \frac{dT}{dx} \ . \tag{3.42}$$

The diffusion approximation fails near the boundaries of a medium even if the medium is nontransparent. Let x_0 be the coordinate of a boundary. In real substances there is always a volume element at a coordinate x' that satisfies $x'-x_0 < \ell_{Rad}$. As a consequence, if the medium is nonconducting, a temperature jump arises near the boundaries that is analogous to the corresponding temperature jump observed for low gas pressures, i.e. large ℓ_{Gas}. Boundary conditions have to be modified in order to apply the diffusion solution also in the immediate vicinity of the wall [Siegel and Howell 1972 pp.639-644]. Naturally, corrections for a temperature jump must vanish if N_1 is very large, that is, if the medium is conductive. As a "conduction model", the diffusion solution is subject to the same rules that deny temperature slips in a medium if it has a "true" conductivity. Figure 3.14 shows dimensionless temperature profiles in a homogeneous medium calculated from the diffusion model solution using a diffusion slip correction [Siegel and Howell 1972 p.642] with temperature independent λ_{Cond} and E as a function of relative optical thickness τ/τ_0. As is seen from the figure, the temperature jump near the walls disappears if τ_0 exceeds 10 or if N_1 equals 10. Note again that the curves do not depend on albedo, because of the additive approximation.

If λ_{Cond} and E depend on temperature, in general there is no closed form solution of $\Theta(\tau)$. We can, however, calculate the temperature from (3.42). Conservation of energy (1.13) requires for all x

$$\nabla_x \cdot \dot{q} = -\frac{d}{dx}\left\{\left[\lambda_{Cond}(x) + \frac{16\sigma \cdot n^2}{3E(x)} \cdot T^3(x)\right] \cdot \frac{dT}{dx}(x)\right\} = 0 \tag{3.43}$$

or, if the integration over dx is carried out, yielding a constant C_1 that equals total heat flow $\dot{q}$

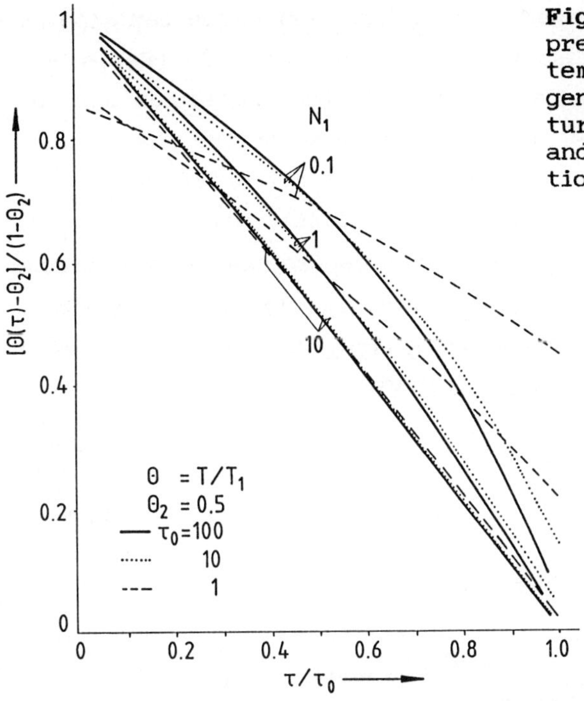

Fig.3.14. Diffusion model predictions for dimensionless temperature profiles in a homogeneous medium using temperature-independent λ_{Cond} and E and a diffusion slip approximation (see text)

$$-\frac{dT}{dx}(x) = \frac{\dot{q}}{\lambda_{Cond}(x) + \dfrac{16\sigma\cdot n^2}{3E(x)}\cdot T^3(x)} . \tag{3.44}$$

This equation enables us to calculate temperature profiles also for temperature dependent $\lambda_{Cond}[T(x)]$ and $E[T(x)]$. Before this is done, another modification of the diffusion model according to anisotropic scattering will be discussed.

3.4.3 Inclusion of Anisotropic Scattering

As in (3.31,32) a correction will be applied to τ_0 that is well known in other fields of research, e.g. in the "transport approximation" for the description of neutron diffusion processes [Glasstone and Edlund 1961 pp.81-83]

$$\dot{q}_{Rad} = -\frac{16\sigma\cdot n^2}{3E\cdot(1-\Omega\cdot\bar{\mu})}\cdot T^3\cdot\frac{dT}{dx} . \tag{3.45}$$

If scattering is conservative ($\Omega = 1$), this correction can also be derived from the usual definition of the K integral [Reiss and Ziegenbein 1985b]. Since all derivations of this correction and the derivations of the LAS models apply an expansion of the phase function, the results naturally coincide. The first-order "diffusion model LAS approximation" is thus given by (3.45), with $\Omega \cdot \bar{\mu} = \omega_1/3$. According to this equation, $\dot{q}_{Rad}$ in an anisotropically scattering medium of optical thickness τ_0 equals the $\dot{q}_{Rad}$ in an isotropically scattering medium of optical thickness $\tau_0 \cdot (1-\omega_1/3)$. The corresponding optical thickness and extinction coefficients

$$\tau_0^* = \tau_0 \cdot (1-\omega_1/3) , \tag{3.46a}$$

$$E^* = E \cdot (1-\omega_1/3) \tag{3.46b}$$

will be called "effective" values of these quantities in the following.

In the case of temperature independent λ_{Cond} and E^* the radiative term in (1.21) simply transforms to

$$\dot{q}_{Rad} = (4/3)\sigma \cdot n^2 \cdot \frac{T_1^4 - T_2^4}{\tau_0 \cdot (1-\omega_1/3)} . \tag{3.47}$$

Apart from the factor n^2 (see below), this equation is identical to (3.32). Yuen's LAS model (3.31) is in a certain sense superior to (3.47) at small τ_0 [9] because Yuen's expression reduces to the familiar result $\dot{q}_{Rad} = \sigma \cdot (T_1^4 - T_2^4)$ for black walls if τ_0 vanishes. For non-transparent media, the term "1" in the denominator of (3.31), i.e. the only difference between the two LAS models, plays no role, of course. Besides Fig.3.11, another comparison of Yuen's expression with exact calculations shows very good agreement [Yuen and Wong 1979]. Accordingly, both LAS expressions are useful representations of $\dot{q}_{Rad}$ in anisotropically scattering media at large optical

[9] This is naturally the case since the description of $\dot{q}_{Rad}$ at small τ_0 is not the task of a diffusion model.

thicknesses. We will report further evidence for (3.45,47) in Chaps.4 and 5. A deeper insight into the reason why (3.31) and (3.45,47) yield reliable predictions of $\dot{q}_{Rad}$ is gained by considering scaling transformations of the equation of transfer [McKellar and Box 1981].[10]

3.4.4 Temperature Profiles Calculated with Temperature-Dependent Parameters

We return now to (3.44), which allows calculation of temperature profiles also in the case of arbitrary temperature dependencies of $\lambda_{Cond}(T)$ or $E(T)$, e.g. if (3.40) applies to $E(T)$. A stepwise calculation of $T(x)$ can always easily be performed. If $x'' > x'$, the simplest approximation is

$$T(x'') = T(x') + (x''-x')\cdot\frac{dT(x')}{dx} \; . \tag{3.48}$$

Since the curvature of temperature profiles in nontransparent media is necessarily small, this approximation is in most cases of sufficient accuracy. The same argument applies to the choice of the starting value $T(x')$ ($0 < x' \ll D$, D: sample thickness). It simply equals wall temperature: if $N_1 > 1$, temperature slip corrections [Siegel and Howell 1972 p.641] are negligibly small compared to τ_0. As a consequence, $T(x') \simeq T(x=0)$. The nontransparent medium behaves like a mirror that reflects wall radiation of the corresponding temperature $T(x=0)$ if τ_0 is sufficiently large.

A very simple case of temperature dependence of $\lambda_{Cond}(T)$ and $E(T)$ has been illustrated by the temperature profiles given in Fig.3.15. It was assumed that the temperature dependencies were given by

$$\lambda_{Cond}(T) = a_1 + b_1\cdot T \; , \tag{3.49}$$

$$E^*(T) = b_2\cdot T \tag{3.50}$$

for reasons that will become obvious in Chap.5 [application of (3.49,50) explains the curvature of $\lambda(T^3)$; using these simple depen-

[10] Scaling of scattering into nonscattering terms can also be applied to transform the transfer equation into an easily integrable differential equation [Lee and Buckius 1983].

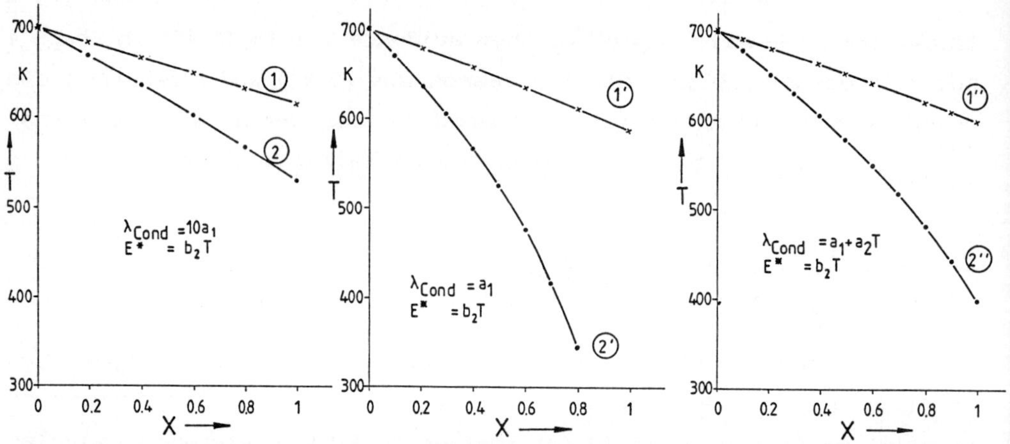

Fig.3.15. Diffusion model predictions for absolute temperature profiles, calculated stepwise by (3.48), in a homogeneous medium, applying temperature dependent λ_{Cond} and E^* (3.49,50), using $\lambda_{Cond} = a_1 + a_2 \cdot T$ with $a_1 = 1$ mW/(m·K), $a_2 = 0.005$ mW/(m·K^2). Curves 1 and 2 refer to $E^* = b_2 \cdot T$ with $b_2 = 5$ or 15 1/(m·K), respectively. All curves use the same $\dot{q} = 100$ W/m^2, $T_1 = 700$ K, $D_0 = 30$ mm. The coordinate x denotes relative geometrical position D/D_0

dencies of λ_{Cond} and E, analytical integration of (3.44) is, of course, possible]. As a result of (3.46b), these temperature profiles depend on albedo Ω but since a demonstration of this is trivial, it has been omitted in Fig.3.15.[11]

A warning should be issued at this point: As the curves 1, 1' and 1" in this figure show, a linear temperature profile does *not* indicate temperature independent λ_{Cond} and E^* (only the reverse is true). Also, a linear temperature profile does *not* imply vanishing radiative transfer (Sect.5.2).

3.4.5 Experimental Determination of Thermal Conductivity Components

Equation (1.22) (with the effective value E^* of extinction coefficient instead of E) can be used to determine E^* as an average taken over the sample thickness from a calorimetric measurement of total λ. If again λ_{Cond} and E^* are independent of temperature, (1.22) yields

[11] It is not desirable to plot dimensionless temperature profiles as in Fig.3.14 because Θ_2 is not constant in Fig.3.15.

$$\lambda = \lambda_{Cond} + \frac{4\sigma \cdot n^2}{3E^*} \cdot T^{*3} = \alpha + \beta \cdot T^{*3} \quad , \tag{3.51}$$

i.e. λ should be linear in T^{*3}. In this equation, T^* is an effective radiation temperature (some researchers prefer the symbol T_{Rad}),

$$T^{*3} = \frac{T_1^4 - T_2^4}{T_1 - T_2} = (T_1^2 + T_2^2) \cdot (T_1 + T_2) = 4T_{Rad}^3 \ . \tag{3.52}$$

A plot of $\lambda(T^{*3})$ thus yields E^* from the slope and λ_{Cond} from the intercept $\lambda(T^{*3}=0)$ of the experimental data. A few examples of the application of this method will now be given.

Figure 3.16a shows total λ for pure silica aerogel in vacuum measured at low and cryogenic temperatures [Kaganer 1969a p.77]. Note the increase of λ_{Cond} and the simultaneous increase of E (i.e., decreasing slope of the curves) with increasing density. If 45 wt.% of a metal powder is added to silica aerogel, λ_{Cond} again increases with increasing density (by compression) but E remains almost constant (Fig.3.16b, [Kaganer 1969a p.97]). Obviously, compression increases the action of the opacifier only gradually since its concentration is already very high. Figure 3.16c shows the total λ of opacified fumed silica in vacuum, under atmospheric pressure load and at high temperatures. Finally, the λ of pure and opacified fibres measured in vacuum and at high temperatures is given in Fig.3.16d. A strong reduction of λ_{Cond} of fibres compared with fine powders is demonstrated by these data and is attributed to a preferential orientation of the fibre axes to temperature gradient, see Fig.4.5 and the discussion of Strong et al. [1960], and Reiss and Ziegenbein [1985c]. It is concluded that the thermal conductivity λ of all these media is in a good approximation linear in T^{*3}, at least within the investigated intervals of this variable (Sects.5.2,3 and Fig.5.6).

Although the measurements were performed at low temperatures, curve 3 in Fig.3.16a shows that at $T^{*3} = 4 \cdot 10^7$ K^3, the radiative contribution to the total λ amounts to more than 90%. From curve b in Fig.3.16c, we have at T_1 = 400°C about 38% radiative contribution, although ℓ_{Rad} was only about 80 μm (total sample thickness about 30 mm, porosity 0.9).

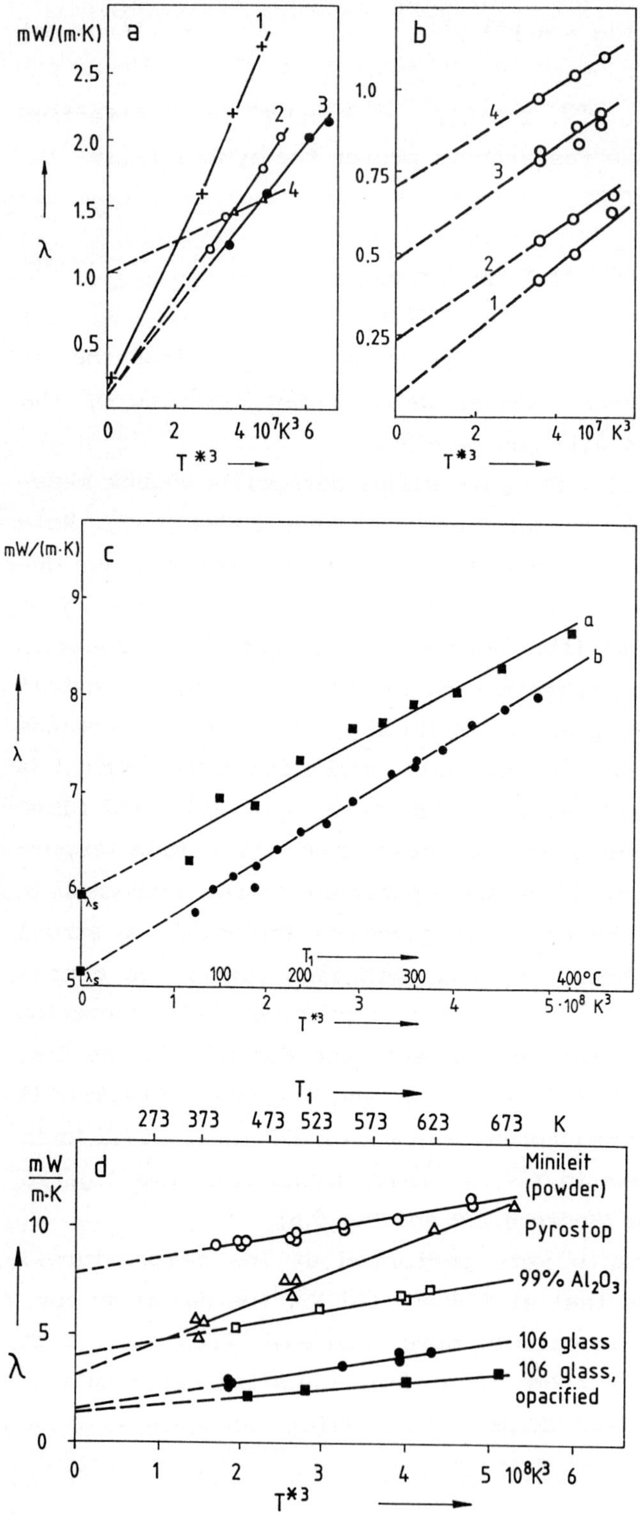

Fig.3.16. Experimental total thermal conductivity of different powders and fibres plotted versus T^{*3} [defined in (3.52)]. Measurements were performed in vacuum and at cryogenic, low and high temperatures.

(a) Pure silica aerogel at densities 70, 105, 125 and 457 kg/m^3, curves 1 to 4, respectively [Kaganer 1969a p.77].

(b) Opacified aerogel (45 weight % bronze powder) at densities 66, 87, 105 and 112 kg/m^3, curves 1 to 4, respectively [Kaganer 1969a p.97].

(c) Opacified fumed silica (curve a: TiO_2 and Fe_3O_4, each 8 weight %; curve b: 17 weight % Fe_3O_4) at a density of 270 kg/m^3. From [Büttner et al. 1983] reprinted by permission of Pion Ltd.

(d) Pure and opacified glass and ceramic fibres at densities between 270 and 350 kg/m^3 [Ziegenbein 1983]. Reprinted by permission of Pion Ltd.

Although this procedure to determine λ_{Cond} and E^* has been used frequently in low temperature physics [Cockett and Molnar 1960; Kaganer and Semenova 1967; Kaganer 1969a], its application to high temperatures is apparently restricted to the experimental work of the present author, his collaborators and those at the University of Würzburg, as described in Sects.4.2 and 5.3.

A procedure to experimentally determine local values of $\lambda_{Cond}(T)$ and $E^*(T)$ is outside the scope of this subsection but will be described in Chap.5.

Another word of caution concerns possible transmission windows at large wavelengths. These can significantly disturb a successful separation of λ_S from λ_{Rad}. Caps [1985] reports a critical wavelength of at least 120 μm up to which spectral extinction coefficients E_Λ have to be large, otherwise an *additional* pseudo-conductivity in the order of 1 mW/(m·K) would be measured. Since scattering cross sections are proportional to $1/\Lambda^4$, a medium is nontransparent at large wavelengths only if it has strong absorption properties in this spectral range.

Extraction of λ_{Cond} from calorimetric measurements has been used successfully to optimize the thermal resistance of insulators (Sect.4.2). However, determination of λ_{Cond} from an extrapolation of measured $\lambda(T^{*3})$ to λ at absolute zero of temperature is in principle unsatisfactory. The question arises as to whether new experiments could be designed to avoid creation of a radiative flow and subsequent elimination of the same quantity from the total heat flow in order to obtain λ_{Cond} as a residual. An interesting correlation between the total thermal conductivity λ of bricks and sound velocity c has been reported (Fig.3.17, [Le Doussal and Bisson 1980]). Since the density of the bricks was high (930 kg/m^3), vanishing radiative flow and thus $\lambda \simeq \lambda_{Cond} \propto c$ was observed in these experiments.

However, sound does not solely consist of periodically propagating displacements with no other physical effects involved. Like the coupling of electric and magnetic fields in electrodynamics, temperature and local displacements of a solid are coupled by a set of thermoacoustic wave equations [Favro et al. 1987]. The solutions of these wave equations (periodic sources assumed) are a longitudinal sound

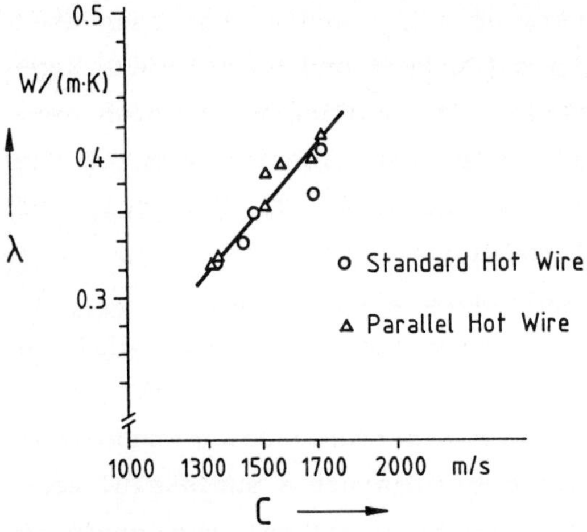

Fig.3.17. Correlation between total thermal conductivity $\lambda = \lambda_{SM}$ of bricks and ultrasonic velocity c [Le Doussal and Bisson 1980]

wave consisting of a combination of local displacements and temperature variations caused by compression of the material, a transverse shear wave that propagates without temperature variations, and a heavily damped thermal wave. Interactions between these different types of waves occur if they encounter a thermal or elastic inhomogeneity. A separation of thermal (i.e. radiation inducing) and acoustic propagation thus seems a priori impossible in dispersed materials, which by definition resemble a *cloud* of inhomogeneities (particle surfaces, solid contacts).

3.4.6 The "Effective Index of Refraction"

The factor n^2 which appears in (1.16,21,22; 3.36,37,42–45,47,51) is the square of the real part of the refractive index. Generally, n^2 modifies the hemispherical emissive power $e_{\Lambda b}$ or e_b of a black body that radiates into a medium other than vacuum [Siegel and Howell 1972 pp.31–32]:

$$e_{\Lambda b,m} = \frac{n^2 \cdot 2\pi \cdot C_1}{\Lambda^5 \cdot \{\exp\left[C_2/(\Lambda \cdot T)\right] - 1\}} = n_m^2 \cdot e_{\Lambda b} \tag{3.53}$$

where the index m indicates the radiation emitted to the medium with refractive index n_m. Caren [1969] claims a factor n^3 instead of n^2 if a metal radiates into a dielectric. The additional factor n stems from an analogous transformation of the thermal emissivity of the metal [that is not accounted for in (3.53)]. It is to be expected that an effective value n_m has to be applied if radiation is incident on a two-phase medium because insertion neither of n_S of a solid phase nor of n_{Gas} of a surrounding gas into (3.53) will approximately account for the composite action of a multiplicity of scattering grain surfaces and a continuous gas on an incoming beam. Frequently, n_m (omitting wavelength index Λ) is calculated as the weighted mean (with respect to volume) of n_S and n_{Gas} (if porosity is high, n_m usually turns out to be close to 1.0). It seems that this practice is not very well founded. Optical constants are not additive, see, e.g., [Volz 1983].

Figure 3.18 shows a comparison between the n^2 that are achieved from a weighted mean of n_S and n_{Gas} (dashed curves) and an expression

$$n_m^2 = \epsilon' = \epsilon_C \cdot \left[1 + \frac{3V_p \cdot \dfrac{\epsilon_S - \epsilon_C}{\epsilon_S + 2\epsilon_C}}{1 - V_p \cdot \dfrac{\epsilon_S - \epsilon_C}{\epsilon_S + 2\epsilon_C}} \right] \tag{3.54}$$

already derived by [Maxwell-Garnett 1904]. In this equation, ϵ' is the average dielectric constant of the two-phase medium ($n^2 = \epsilon$

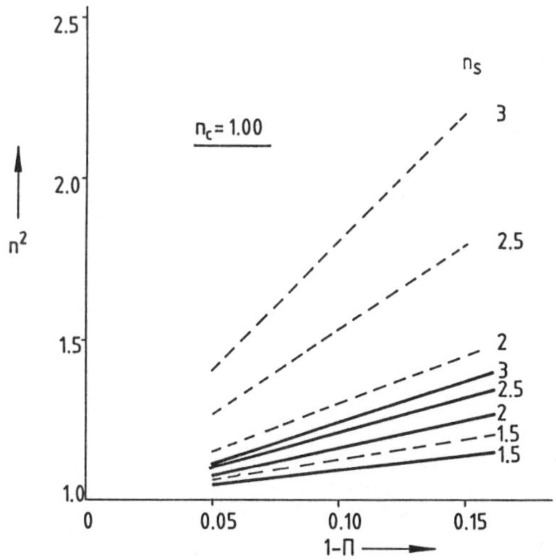

Fig.3.18. Square of the "effective" refractive index n^2 of spherical dielectric particles (refractive index n_S) embedded in a continuum (refractive index n_C), calculated using (3.54) (solid curves) and as a weighted mean (with respect to volume, dashed curves). Π denotes porosity

according to Maxwell's relation), $\epsilon_S = n_S^2$, e_C denotes the ϵ of the continuous phase (i.e., the ϵ of the gas), and V_p is the relative volume contribution of the solid phase. Equation (3.54) uses a derivation of a mean value of ϵ' as a function of the Lorentz-Lorenz polarizability of spheres. Note that (3.54) is related to the well-known Clausius-Mosotti equation: according to Clausius and Mosotti, for any given substance, for pure electronic polarizability, for a cubic lattice and in the absence of permanent dipoles, $(\epsilon-1)/(\epsilon+2)$ should be proportional to the density and the polarizability of a substance [Jackson 1967 p.119]. However, (3.54) is used in a different sense: an aggregate of small spheres made from a material of refractive index n_S that are distributed in a continuous phase, is optically equivalent to a medium of refractive index $n^2 = \epsilon'$.

According to Fig.3.18, the frequently observed simplified calculation of n^2 as a weighted mean is certainly an overestimate.

It is further obvious that a wavelength average $\langle n^2 \rangle$ of spectral values $n^2_{m,\Lambda}$ calculated from (3.54) has to be inserted into the expressions for $\dot{q}_{Rad}$ (a proof of this is simple and straightforward). The $n^2_{m,\Lambda}$ calculated from (3.54) are very close to 1 as long as $n \leq 3$ and $\Pi > 0.9$. Only slight alterations to $e_{\Lambda b}$ have to be expected in these dispersed media. In metal powders, however, the corrections can become important at large wavelengths where n and k reach values in the order of 100. The same applies to other strongly absorbing dispersed substances (Sect.6.3).

4. Comparison Between Measured (Calorimetric or Spectroscopic) and Calculated Extinction Coefficients

4.1 Materials Used in Calorimetric and Spectroscopic Experiments

Most of the investigations described in this chapter have been performed with pure dielectric powders or fibres or using mixtures with infrared optical opacifiers that use dielectric powders or fibres as a skeleton. Note that the classification into absorbing or scattering opacifiers is largely determined by the particle diameters in the substances.[1] It is not possible to deduce scattering or absorption properties exclusively from magnitudes of real and imaginary part of the refractive index. If particle diameter increases, scattering will eventually dominate because of increasing diffraction (Fig.4.1 or [Yurevich and Konyukh 1975]). As a consequence, substances that have large electrical conductivity, i.e. large imaginary part of refractive index, can also be considered scattering opacifiers at appropriate diameters and/or wavelengths.

Figure 4.2 shows Fe_3O_4 powder which has been applied frequently as an opacifying agent in these investigations. A mixture of fumed silica with Fe_3O_4 is analyzed by SEM micrographs in Fig.4.3a,b. Figure 4.4a,b shows the same analysis applied to a mixture of fumed silica with $FeTiO_3$. Figure 4.5 gives a SEM micrograph of a glass fibre paper that was used predominantly for the measurement and analysis of temperature profiles in fibrous media discussed in Chap.5. Note that application of glass fibre paper allows a preferential orientation of the fibres perpendicular to temperature gradient. When designing

[1] Whether scattering or absorption is dominant in different opacifiers has been clarified by Trunzer [1983]. This reference also contains measured specific BET surfaces.

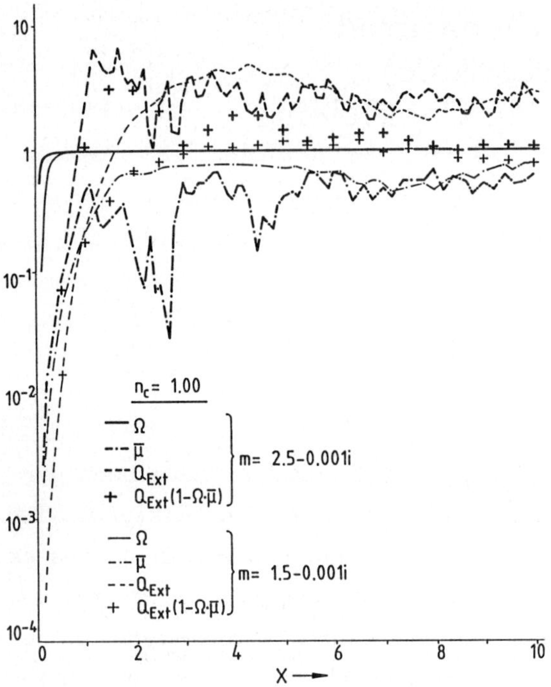

Fig.4.1. Calculation of albedo Ω, asymmetry factor $\bar{\mu}$ (2.66), relative extinction cross section Q_{Ext} and effective values $Q_{Ext}(1-\Omega\cdot\bar{\mu})$ from rigorous Mie theory for spherical dielectric particles as a function of scattering parameter $x = \pi\cdot d/\Lambda$ for different complex refractive indices m

Fig.4.2. Scanning electron micrograph of Fe_3O_4 powder; the length of the horizontal bar denotes 1 μm

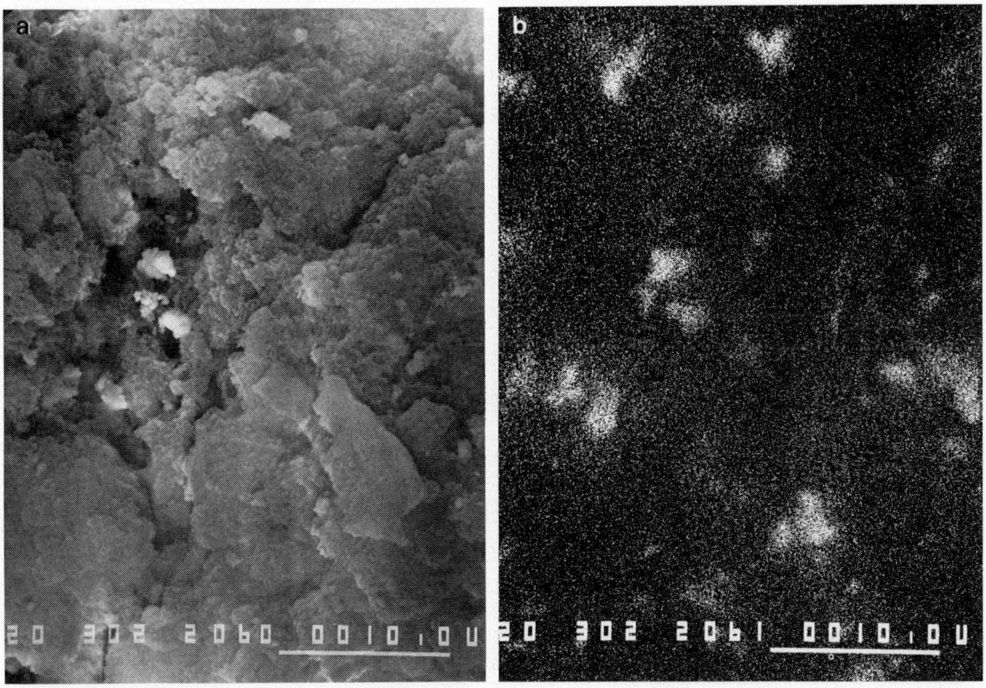

Fig.4.3. Scanning electron micrograph (a) and Fe-Kα intensities (b) measured in a μ-probe of a mixture of fumed silica with Fe_3O_4. Single particle diameter of fumed silica or Fe_3O_4 was about 0.007 μm or 0.2 μm, respectively; the length of the horizontal bar denotes 10 μm. [Reiss 1983] copyright AIAA (1981)

thermal insulations, it is well known that a strong increase of heat flow occurs if the fibres are not oriented in this manner [Strong et al. 1960; Tye and Desjarlais 1983]. Similar observations have been made when the thermal conductivity of stratified solids like graphite or mica was measured under different orientations [Cabannes 1980a].

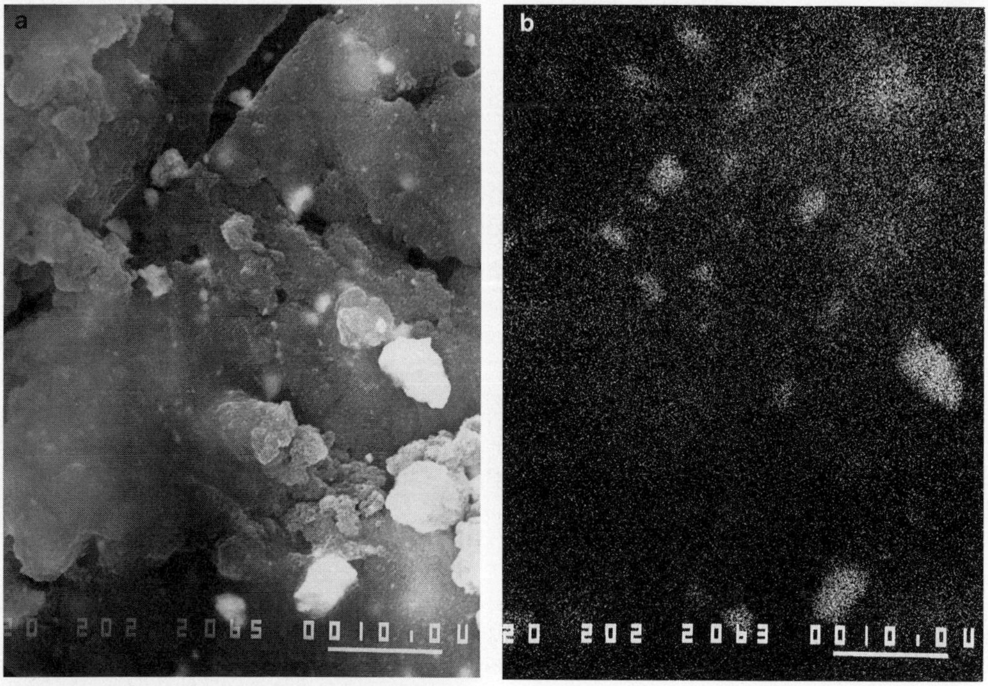

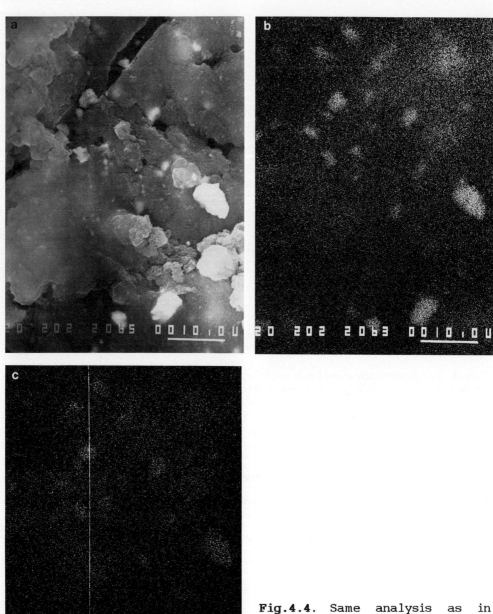

Fig.4.4. Same analysis as in Fig.4.3 of a mixture of fumed silica with $FeTiO_3$. (a) Electron micrograph. (b) Ti-Kα intensities. (c) Fe-Kα intensities. The length of the horizontal bar denotes 10 μm. [Reiss 1983] Copyright AIAA (1981)

Fig.4.5. Scanning electron micrograph of borosilicate glass fibre paper. Note the preferential orientation of the fibres [Reiss and Ziegenbein 1983]. The length of the horizontal bar denotes 100 μm

4.2 Calorimetric Measurements

Measurements of thermal conductivity of dispersed media that were performed under clearly defined mechanical boundary conditions are rather scarce [Strong et al. 1960; Büttner et al. 1983, 1985a,b; various papers of this author since 1981 (see list of references)]. A large guarded hot plate device (Fig.4.6, original design) has been used to investigate thermal conductivity of load bearing powders and fibres in vacuum at high temperatures and under constant atmospheric or (after alterations) variable pressure load.

Figure 4.7 shows total thermal conductivity λ of fumed silica plus Fe_3O_4 (curve a), pure glass fibres (curve b), and glass fibres plus Fe_3O_4 (curve c). All measurements [Büttner et al. 1985a] were performed in vacuum, under atmospheric pressure load and at high temperatures. Data are plotted versus T^3_{Rad}, which is defined in (3.52).

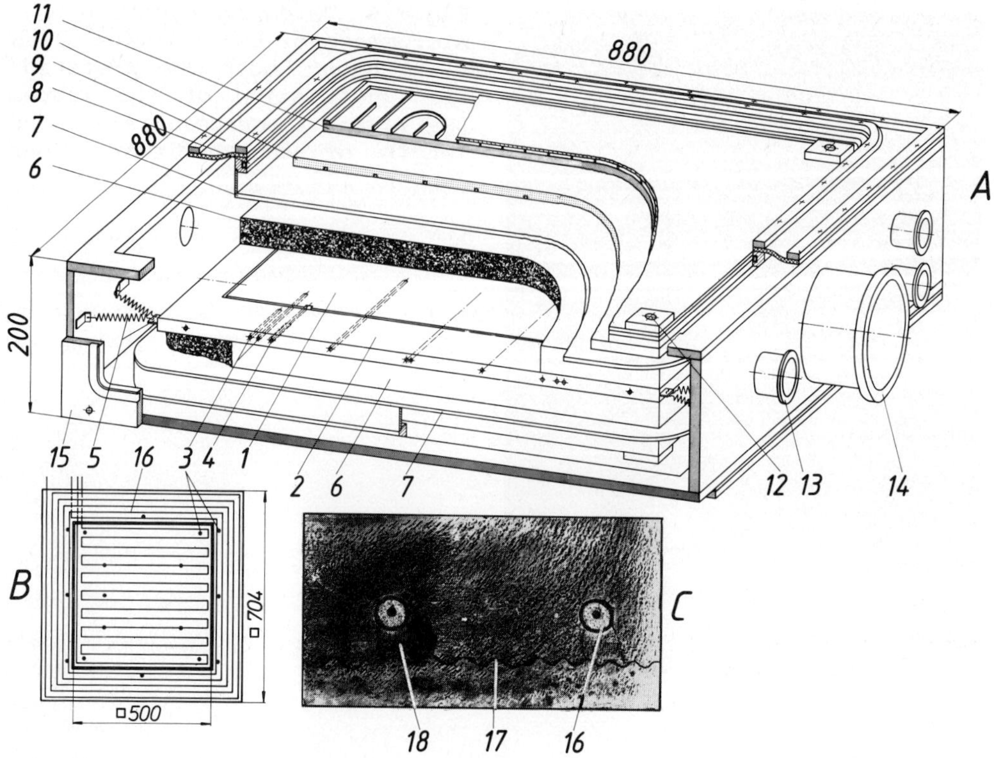

Fig.4.6. Guarded hot plate device for measurement of thermal conductivity in vacuum and under external atmospheric pressure load. (A) Overall view. (B) Location of temperature sensors (dots) in heating plate. (C) Cross section of explosion-welded joint showing heating coil. For details see [Büttner et al. 1983] reprinted by permission of Pion Ltd.

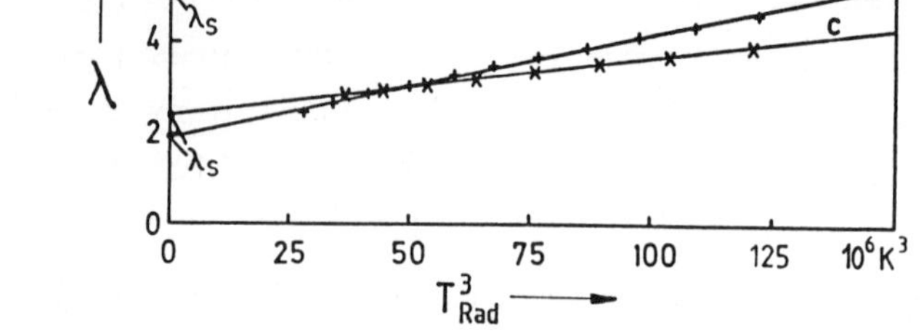

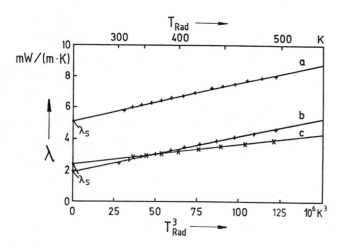

Fig.4.7. Total thermal conductivity λ of an opacified powder insulation (a), a pure (b) and an opacified (c) glass fibre insulation plotted versus $T_{Rad}{}^3 = T^{*3}/4$. [Büttner et al. 1985a] reprinted by permission of Pion Ltd.

Table 4.1 collects temperature independent, thickness averaged, solid thermal conductivity contributions $\lambda_S = \lambda_{Cond}$ to the total λ that have been extracted at $\lambda(T^{*3}=0)$ from different experimental work performed in vacuum and at low and high temperatures. λ_S of pure fibres that were oriented perpendicular to temperature gradient are considerably lower compared to the λ_S of powders. The relatively high λ_S of powders is probably due to the high pressure (about 0.5 to 0.7 MPa) required to prepare plates from the mixtures (curve a in Fig.4.7; a small quantity of glass fibres was added to the powders to increase the bending strenght of the plates). The dependence of λ_S on concentration of Fe_3O_4 in fumed silica is rather strong [Reiss 1981a,b]. This is certainly due to numerous Fe_3O_4 particle-particle contacts and Fe_3O_4 clusters that arise if the concentration of this component is high (see Fig.4.3a,b). On the other hand, isolated Fe_3O_4 particles that are attached to well-separated fibre surfaces will hardly contribute to λ_S. This is confirmed by comparing λ_S of curves b and c in Fig.4.7.

From the slope of the curves a, b, and c in Fig.4.7 and from other experimental work (temperature independent) thickness averaged, specific extinction coefficients E^*/ρ given in Table 4.2 are extracted [for the definition of E^* see (3.46), and ρ denotes density]. The E^*/ρ of pure glass fibres are about the same as those of heavily doped fumed silica (the well-known transparency of SiO_2 is reflected by these results).

Still higher calorimetric E^*/ρ than those given in Table 4.2 are reported by Grunert et al. [1969a] for glass fibre paper opacified with aluminum flakes, $E^*/\rho \geq 0.26$ m^2/g. However, no comparison with spectroscopy or theory was presented.

Using the extinction coefficients of Table 4.2 together with (3.51), a determination of λ_{Rad} from calorimetric, spectroscopic or theoretical work is immediately possible.

An important test of the validity of the diffusion model applying temperature independent, thickness averaged λ_S and E^* is a measurement of $\lambda(T^{*3})$ using the wall temperatures $T_i, T_j \neq T_k, T_\ell$. Only if $\lambda(T_{ij}{}^{*3}) = \lambda(T_{k\ell}{}^{*3})$ is experimentally confirmed for $T_{ij}{}^{*3} = (T_i^2+T_j^2)$

Table 4.1: Conduction components λ_{Cond} extracted from $\lambda(T^{*3})$-plots of different powders and fibres in vacuum.

Material	Density [kg/m^3]	λ_{Cond} [mW/(m·K)]	Reference
Silica aerogel	70	0.14	Kaganer 1969a p.77
Silica aerogel (Santocel A)		0.06	Cockett and Molnar 1960
Silica aerogel	457	1.0	Kaganer 1969a p.77
Silica aerogel + 45 wt.% bronze powder	87	0.23	Kaganer 1969a p.97
Silica aerogel + 45 wt.% bronze powder	105	0.48	"
Silica aerogel + 45 wt.% bronze powder	112	0.7	"
Silica aerogel + 29% aluminum powder		0.14	Kaganer 1969a p.92
Silica aerogel + 40% aluminum powder		0.18	"
Silica aerogel + 50% aluminum powder		0.18	"
Santocel A + 40% aluminum powder		0.13	Cockett and Molnar 1960
Calcium silicate (Microcel E)		0.35	"
Soot, d = 0.1 μm	250	4.0	Serebryanyi et al. 1968
Cement, d = 5 μm	1500	4.0	"
Perlite	100	0.1	Kaganer 1969a p.78
Perlite	150	0.16	"
Perlite	360	0.8	"
Fumed silica + 16 wt.% Fe_3O_4	270	5.10 *a*	Büttner et al. 1983
Fumed silica + 8 wt.% TiO_2 + 8 wt.% Fe_3O_4	270	5.9 *a*	"
Minileit (fumed silica + $FeTiO_3$, Grünzweig & Hartmann AG, Ludwigshafen)	260	7.8 *a*	Ziegenbein 1983
Pure borosilicate glass fibres	300	1.90 *a*	Büttner et al.1985a
Ceramic fibres (99% Al_2O_3, Didier AG, Wiesbaden)	330	4.00 *a*	Ziegenbein 1983
Borosilicate glass fibres + 30 wt.% Fe_3O_4	330	2.40 *a*	Büttner et al. 1985a
Aluminized glass fibres	57	0.44	Reiss et al. 1985

a Measurements performed in vacuum and under atmospheric pressure load.

Table 4.2: Comparison of temperature independent E^*[1/m] or E^*/ρ[m^2/g] of nontransparent dispersed media from calorimetric and spectroscopic measurements with prediction of Mie theory (all values assuming $n^2=1$; percentages of opacifiers by weight).

Material	Integral Extinction		
	Calorimetric	Spectroscopic	Theory
Perlite	$E^* = 3.1 \cdot 10^3$ *a*	$3.2 \cdot 10^3$ *a*	
Silica aerogel	$E^* = 2.5 \cdot 10^3$ *a*	$3 \cdot 10^3$ *a*	
Silica aerogel + 20% bronze powder	$E^* = 5.3 \cdot 10^3$ *a*	$4.5 \cdot 10^3$ *a*	
Silica aerogel + 40% bronze powder	$E^* = 1.3 \cdot 10^4$ *a*	$9.4 \cdot 10^3$ *a*	
Fumed silica + 16% Fe_3O_4	$E^*/\rho = 0.046$ *b*	0.042 *c*	
Pure borosilicate glass fibres	$E^*/\rho = 0.058$ *b*	0.06 *d*	0.06 *d*
Fibreglass bonded mats, d = 14 µm	$E^*/\rho = 0.011$ *e*	0.0087 *e*	
Fibral alumina fibres, d = 5–10 µm	$E^*/\rho = 0.019$ *f*	0.014 *f*	
Borosilicate glass fibres + 30% Fe_3O_4	$E^*/\rho = 0.077$ *b*		~0.1 *i*
Borosilicate glass fibres + 33% Fe_3O_4	$E/\rho = 0.067$ *g*	0.058 *g*	
Borosilicate glass fibres + 50% Fe_3O_4	$E = 2.8 \cdot 10^4$ *g*	$2.62 \cdot 10^4$ *g*	
Aluminized glass fibres	$E^*/\rho = 0.11$ *h*	0.2 *h*	0.24 *h*

a Mil'man and Kaganer 1975; measurement at low temperatures.

b Büttner et al. 1985a.

c Caps et al. 1983a; experimental value E/ρ of Table 4.3 for 17% Fe_3O_4.

d Caps et al. 1984.

e Cabannes et al. 1979.

f Cabannes 1980b.

g Ziegenbein 1983; because of the large amount of Fe_3O_4 these mixtures have dominating absorption properties ($\Omega \simeq 0.33$), i.e. nearly isotropic scattering, so that the given E or E/ρ are close to E^* or E^*/ρ, respectively.

h Reiss et al. 1985.

i Caps 1985.

Calorimetric experiments for *b*, *e*, *f* and *g* were performed at high sample temperatures; in cases *e* to *g*, the authors assumed isotropic scattering; in cases *e* and *f* measurements were not performed in vacuum.

$\cdot(T_i+T_j) = (T_k^2+T_\ell^2)\cdot(T_k+T_\ell) = T_{k\ell}{}^{*3}$, then T^{*3} is a "good" variable, i.e. (λ,T^{*3}) diagrams that have been used extensively in low temperature physics are reproducible. Figure 4.8 shows that this requirement is indeed confirmed when fumed silica plus Fe_3O_4 (curve a in Fig.4.7) was investigated. Since this measurement was performed at different mean temperatures $[T_{ij} = (T_i+T_j)/2 \neq T_{k\ell} = (T_k+T_\ell)/2]$, in turn it can be concluded that thickness averaged λ_S and E^*/ρ of this sample are temperature independent in the temperature interval under investigation.

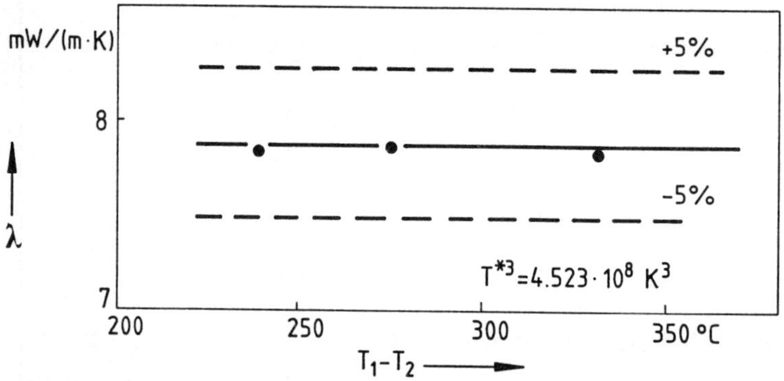

Fig.4.8. An experimental check of the validity of T^{*3} (or T^3_{Rad} as used in Fig.4.7) as a "good" variable (see text). [Büttner et al. 1983] reprinted by permission of Pion Ltd.

Measurements of λ of glass fibres in vacuum and under a *variable* external pressure load p are reported by Büttner et al. [1985a,b]. These measurements allow extraction of $\lambda_S(p)$ and $\lambda_{Rad}(p)$ and comparison with theoretical models. Fig.4.9 shows total k, k_S and k_{Rad} measured between $2\cdot10^3$ Pa $\leq p \leq 10^5$ Pa.[2] The temperature dependent component k_{Rad} is extracted from a fit $k = k(T^{*3},p)$, and $k_S(T^{*3},p) = k(T^{*3},p) - k_{Rad}(T^{*3},p)$.

Although sample thickness D is a function of load p, τ_0 is independent of p because the number of particles does not change (modifications can occur only if dependent scattering arises). Therefore, k_{Rad} must be independent of p whereas λ_{Rad} (by definition of λ) is a function of p, which was confirmed in this experiment.

[2] The definition of k is $k = \lambda/D$ (D: sample thickness).

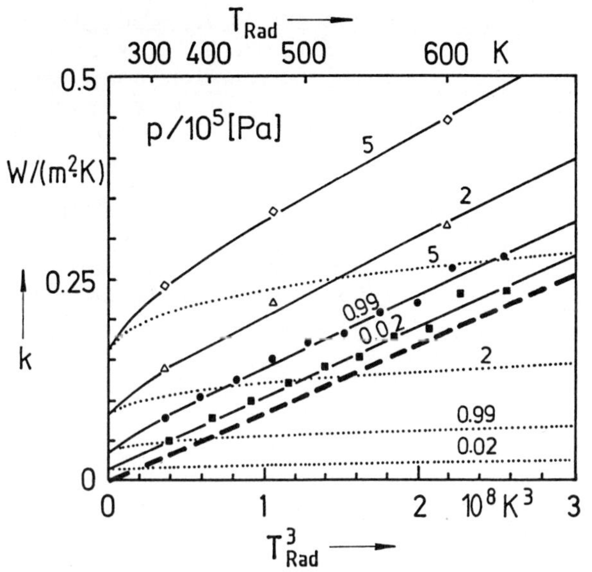

Fig.4.9. Total heat loss coefficient k (symbols and solid curves) and its components k_S (dotted curves) dependent on p and k_{Rad} (dashed curve), independent of p, measured at different pressure load p using borosilicate glass fibre paper in vacuum. [Büttner et al. 1985b] Copyright AIAA (1985)

When extracting λ_S from k_S it was found that $\lambda_S(p)$ can be fitted well below $p=10^5$ Pa by the relation $\lambda_S = \lambda_{S,0} + a\cdot p^{1/3}$, see Fig.4.10 [Büttner et al. 1985a] that follows from Kaganer's model (2.29). Apparently Büttner's work is the first analysis of a *separated* solid thermal conduction component. Cunnington and Tien [1972] investigated a $\lambda(p^{1/3})$ dependence of microspheres, i.e. the total λ was applied in the analysis which is certainly not justified since λ_{Rad} depends on p [Büttner et al. 1985a,b].

Although a serious disadvantage of Kaganer's model is its rigidity (i.e., it contains no free parameter that would allow a description of a non-ideal network), it seems that it predicts the p dependence of λ_S correctly, at least for $p \leq 10^5$ Pa.

4.3 Spectroscopic Measurements

Only an indication of the efficiency of different opacifying agents for attenuation of infrared radiation can be given by integral (i.e. wavelength independent) measurements (Fig.4.11 [Caps et al. 1983a]).

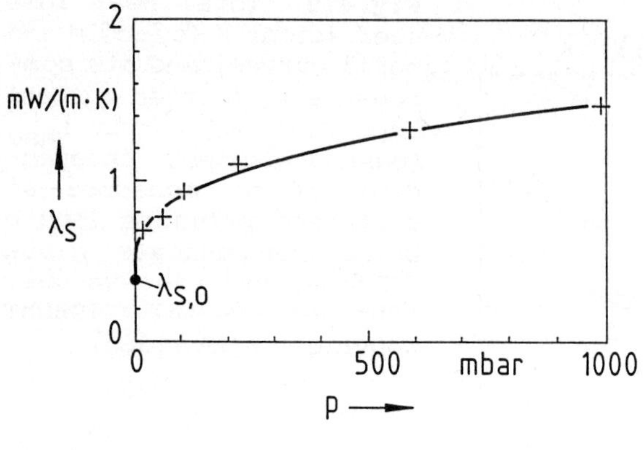

Fig.4.10. Analysis of an experimentally separated solid conduction component λ_S of total λ extracted from Fig.4.9 in terms of a simple resistor model. The solid curve denotes a fit $\lambda_S = \lambda_{S,0} + a \cdot p^{1/3}$ [see (2.29)] to the data (crosses). [Büttner et al. 1985a] reprinted by permission of Pion Ltd.

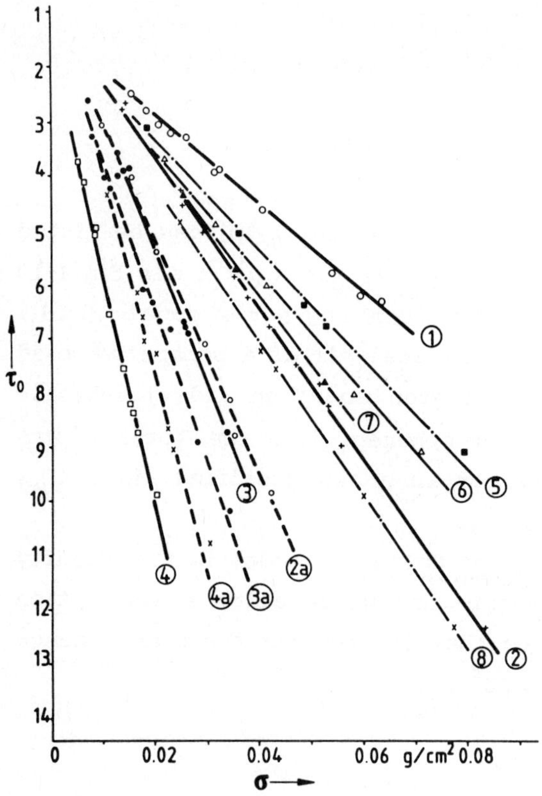

Fig.4.11. Experimental integral optical thickness τ_0 plotted versus surface density σ for different powders and opacifiers; the symbols are identified in Table 4.3. [Caps et al. 1983a] reprinted by permission of Pion Ltd.

Optical thickness τ_0 is plotted versus surface density σ.[3] The slope $d\tau_0/d\sigma$ equals E/ρ and is collected together with other experimental

[3] These experiments use Beer's law. If no scattered radiation falls on the detector, E (not E^*) is measured.

work in Tab.4.3. Obviously Fe_3O_4 and SiC are very effective additives. The role that the opacifiers really play in mixtures with transparent substances can, however, be understood only from spectral measurements of both skeletons and additives.

Table 4.3: Integral extinction coefficients E or E/ρ of various powders and fibres; mixtures of (A) KBr or (B) borosilicate glass fibres with different opacifiers given in weight-percent.

Material	Extinction	No. given in Fig.4.11
(A) **POWDERS** [Caps et al. 1983a]	$E/\rho[m^2/g]$	
Pure fumed silica	0.0081	1
1.33% Fe_2O_3 in KBr	0.010	5
1.27% Cr_2O_3	0.011	6
1.11% SiC	0.013	7
1% Fe_3O_4	0.014	2
2.66% MgO	0.014	8
4.85% $FeTiO_3$	0.021	2a
5% Fe_3O_4	0.024	3
9.38% $FeTiO_3$	0.028	3a
13.23% $FeTiO_3$	0.037	4a
17% Fe_3O_4	0.042	4
(B) **FIBRES** [Ziegenbein 1983]	$E[1/m]$	
Pure borosilicate glass fibres	$1.26 \cdot 10^4$	
33.3% Fe_3O_4 in fibres	$1.93 \cdot 10^4$	
50% Fe_3O_4	$2.62 \cdot 10^4$	
66.7% Fe_3O_4	$3.99 \cdot 10^4$	

All experiments performed at high temperatures of incident black body radiation; samples at room temperature.

Figure 4.12 shows measured spectral E_Λ/ρ of the opacifiers Fe_3O_4 and $MoTe_2$ and of Fe spheres [Caps et al. 1984]. Both curves again demonstrate that E_Λ/ρ decreases with decreasing scattering parameter x (2.57). Although the absorption properties of Fe_3O_4 are strong at low Λ, use of this substance as an opacifier at longer wavelengths

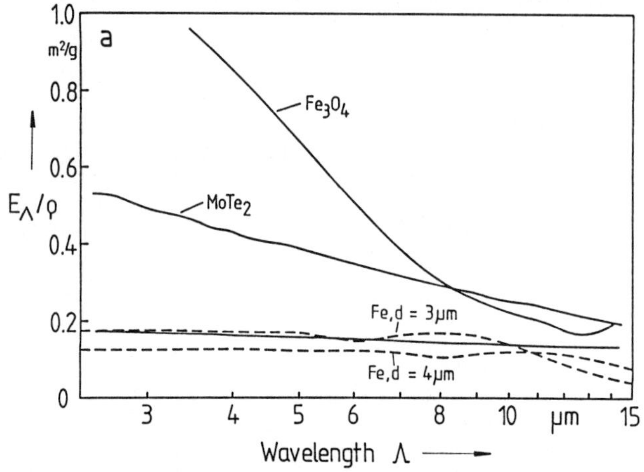

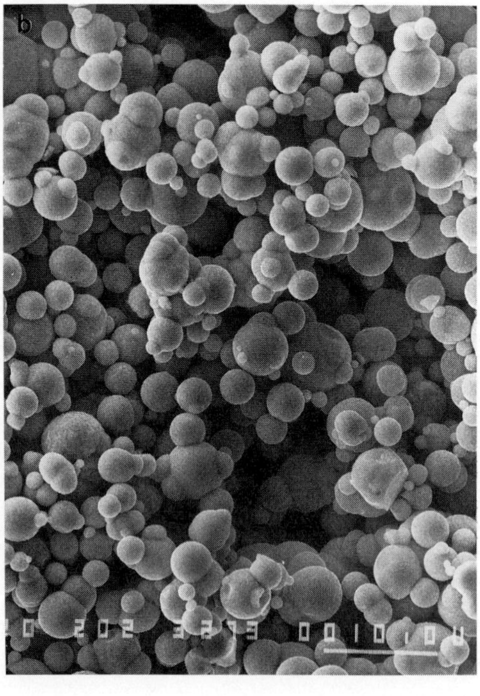

Fig.4.12. Experimental extinction spectra (a, solid curves) of Fe_3O_4, $MoTe_2$ and Fe particles. The dashed curves are calculations of E_Λ/ρ for the Fe particles (see b) using rigorous Mie theory [Caps et al. 1984]. The length of the horizontal bar in (b) denotes 10 μm. Reprinted by permission of Pergamon Journals Ltd.

will be efficient only if the particle diameter is increased. That E_Λ/ρ of Fe spheres is independent of wavelength is in agreement with theory.

Caps et al. [1983b] measured E_Λ/ρ of Microglass 1000 (Manning Corp., Troy, NY) at wavelenths 2 μm $\leq \Lambda \leq$ 15 μm (Fig.4.13). Between 3 μm $\leq \Lambda \leq$ 6 μm, the decrease of E/ρ is again due to a decrease of the scattering parameter x. The strong absorption band at 9.5 μm is characteristic for all inorganic glasses and results from vibrational modes of O-Si-O molecules. The manufacturers claim a mean fibre diameter d of 0.5 μm. However, REM micrographs of the sample indicated that d between 0.5 and 1 μm is more adequate [Caps et al. 1983b].

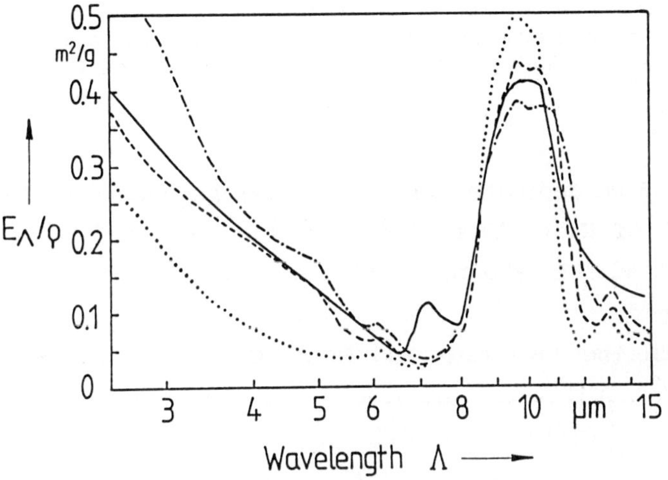

Fig.4.13. Experimental extinction spectrum (solid curve) of borosilicate glass fibres (Microglass 1000, σ = 14.2 g/cm^2) and calculated E_Λ/ρ using rigorous Mie theory and fibre diameters d = 0.5 μm (dotted curve), d = 1 μm (dashed curve) and d = 1.5 μm (dashed-dotted curve) [Caps et al. 1983b]

Figure 4.14 shows experimental reflectances R (and calculations using Mie theory and five-flux model) determined at a low and a very high surface density σ of Microglass 1000 fibres. The $\sigma \to \infty$ corresponds to an infinite optical thickness so that the reflectance R equals R_∞ defined in (3.13) (Fig.3.3b). Both experimental and theoret-

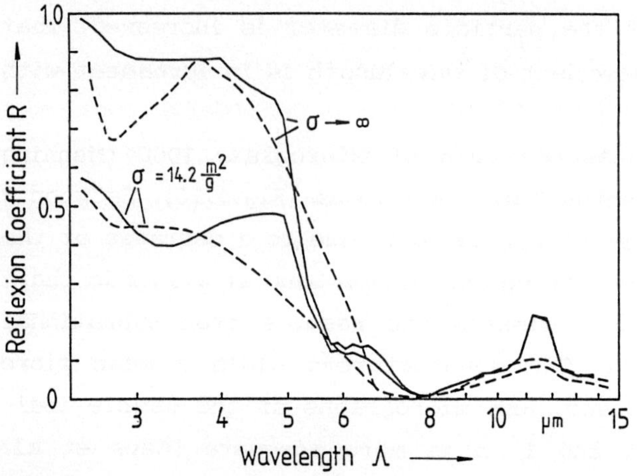

Fig.4.14. Experimental (solid curves) and calculated (dashed curves) reflection coefficients R for different surface densities σ of cold borosilicate glass fibres (Microglass 1000) plotted versus wavelength Λ. The calculations apply predictions from Mie theory (for E and Ω) using particle diameter d = 1.4 μm and five-flux approximations for R. [Caps et al. 1984], reprinted by permission of Pergamon Journals Ltd.

ical R_∞ at 2 μm $\le \Lambda \le$ 5 μm indicate that scattering dominates in the extinction coefficient of Microglass fibres in this region of wavelengths (in Sect.4.4 it will be shown, in addition, that this scattering is highly anisotropic).

For comparison with the five-flux and Monte-Carlo calculations shown in Fig.3.5 for transmittance T at several wavelengths, Fig.4.15

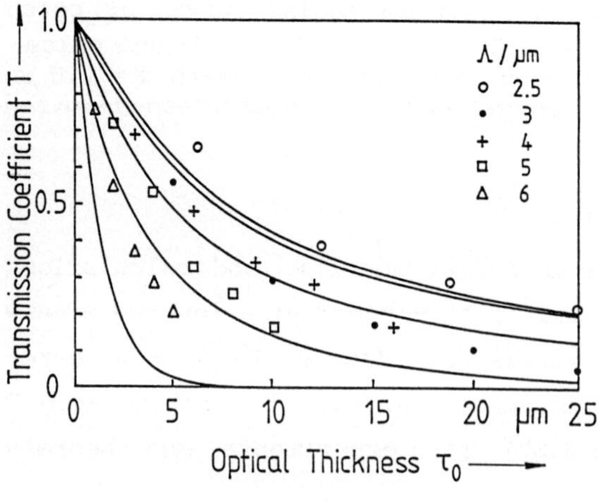

Fig.4.15. Experimental transmission coefficients $T(\tau_0)$ (symbols) of borosilicate glass fibres (Microglass 1000) and calculated $T(\tau_0)$ (solid curves) from five-flux cold medium approximations using particle diameter d = 1.5 μm, predictions of Ω and μ from Mie theory, and experimental optical thicknesses τ_0. [Caps et al. 1984], reprinted by permission of Pergamon Journals Ltd.

gives experimental values for T for glass fibres (Microglass 1000) measured at the same wavelengths. From both Figs.3.5 and 4.15 it is seen that T decreases with increasing τ_0 rather slowly if scattering prevails. For simplicity, the two-flux model (3.11) predicts $T \rightarrow 1/(1+\tau_0)$ if $\Omega \rightarrow 1$, whereas $T \rightarrow \exp(-2\tau_0)$ if $\Omega \rightarrow 0$ [Kaganer 1969a p.47], assuming isotropic scattering. It is seen from Fig.4.15 that the decrease of T is stronger for the larger wavelengths, that is if absorption begins to dominate. Instead of the weak $T \propto 1/\tau_0$ law, an exponential decay of T then causes the observed strong decrease of the transmittance.

4.4 Calculated Extinction Coefficients

A recently performed test of the validity of E^* is reported by Glicksman et al. [1987] who compare hemispherically measured extinction coefficients with scaled (i.e., calculated) ones. The researchers found agreement to within 10% if polyurethane foams were investigated. Scaling the extinction coefficient by the $\ell = 1$ approximation and comparison of diffusion model predictions with exact (numerical) calculations yields agreement within 1% to 3% [however, Glicksman et al. derived extinction coefficients from the slope of the line of logarithm transmissivity versus sample thickness, which yields accurate results only if absorption clearly dominates; see end of Sec.4.3 or Sec.3.1: if $\Omega \rightarrow 0$ then $T(D) \rightarrow \exp(-2E \cdot D)$ in the two-flux approximation].

Rigorous Mie theory has been applied to calculate theoretical E_Λ/ρ of glass fibres for comparison with experiment [Caps et al. 1983b]. The theoretical curves in Fig.4.13 show that excellent agreement with measured E_Λ/ρ of Microglass 1000 is achieved if particle diameter $d = 1\ \mu m$ and complex refractive indices of silicate glass [Hsieh and Su 1979] entered the calculations.

Very good agreement with experiment is also found when the experimental E_Λ/ρ of borosilicate glass fibres are compared either with rigorous Mie theory or with results of an approximate expression

(Fig.4.16 [Reiss and Ziegenbein 1983]). This expression uses anomalous diffraction theory: van de Hulst [1981 pp.312-313] derived for the relative extinction cross sections Q_{Ext} of a transparent (i.e., dielectric) medium the Rayleigh-Gans limit (also [Stephens 1984])

$$Q_{Ext,\Lambda}(\eta_\Lambda) = (2/3)\eta_\Lambda^2 \qquad (4.1)$$

where $\eta_\Lambda = 2x\cdot(n_\Lambda - 1)$. Properly speaking, this expression is valid only for small x and for real refractive indices that are close to 1.

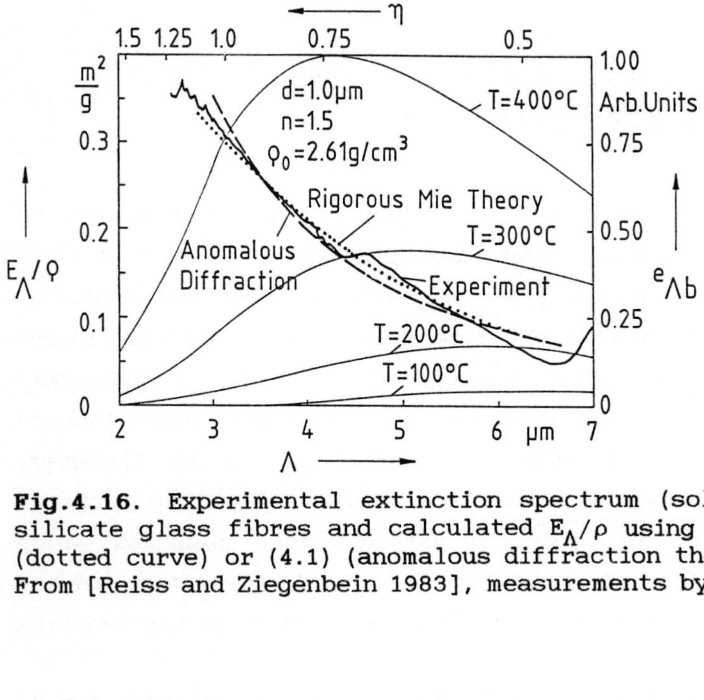

Fig.4.16. Experimental extinction spectrum (solid curve) of borosilicate glass fibres and calculated E_Λ/ρ using rigorous Mie theory (dotted curve) or (4.1) (anomalous diffraction theory, dashed curve). From [Reiss and Ziegenbein 1983], measurements by Dr. R. Caps

The importance of anisotropic forward scattering is revealed if theoretical E_Λ/ρ and E_Λ^*/ρ values for glass fibres following from rigorous Mie theory are compared (Fig.4.17). That the albedo Ω is large at $1\ \mu m \leq \Lambda \leq 5\ \mu m$, i.e. scattering prevails, is in accordance with experimental observations for glass fibres as shown in Fig.4.14; this is also found for ZrO_2 fibres [Cabannes et al. 1979]. Figure 4.17 demonstrates that this scattering is highly anisotropic because the large E_Λ/ρ are drastically reduced to nearly constant E^*/ρ at these wavelengths. As a consequence, an extinction measurement can be

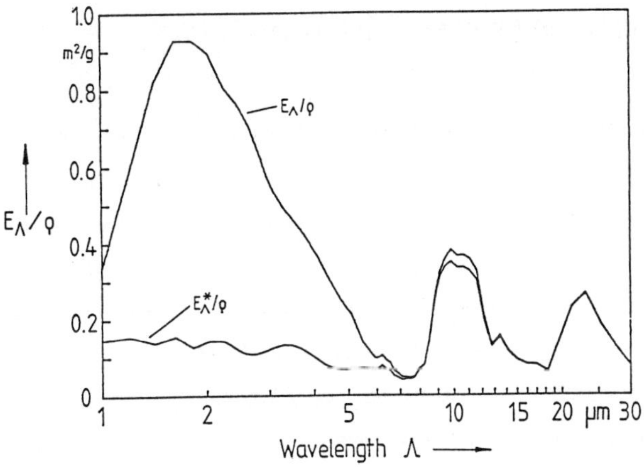

Fig.4.17. Calculated extinction spectra E_Λ/ρ and $E_\Lambda{}^*/\rho = (E_\Lambda/\rho)\cdot(1-\Omega_\Lambda\cdot\mu_\Lambda)$ of glass fibres showing the effect of strong anisotropic (forward) scattering at $\Lambda \leq 7$ μm on the extinction coefficient [Caps et al. 1983b]

largely in error if forward scattered radiation enters the detector. (In astrophysics, strong forward scattering can simulate regions of space with apparently increased star densities).

From the theoretical E_Λ/ρ, the Rosseland mean $E_R{}^*/\rho$ has been calculated for comparison with the two experimental methods (Table 4.2).

4.5 Comparison of Extinction Coefficients Obtained from the Three Independent Methods

Table 4.2 finally collects extinction coefficients that have been obtained from calorimetric and spectroscopic measurements and from application of Mie's theory to different experimental work. The agreement is rather good. Thus it has been shown that a reliable determination of E^*/ρ is possible by three completely independent methods.

Based on these encouraging results, predictions of E^*/ρ for a wide range of particle diameters and temperatures have been made [Caps et al. 1983b]. A special application of Mie theory to very thin fibres of strongly reflecting substances is reported in Chap.6.

4.6 Translucence in a Nontransparent Medium

Although anisotropic scattering in glass fibres at small wavelengths is strong, the Monte Carlo simulation presented in Fig.3.2 demonstrates that after a few optical thicknesses, anisotropy of the radiation field intensity almost completely disappears. This has an important consequence: it is not possible to deduce the degree of anisotropic scattering from an observation of intensities that escape from a sample even if its optical thickness is only about 15 to 20.

A similar result is already known in astronomy: the intensity i' of the Sun's atmosphere is given [Unsöld 1968 p.130] by

$$i'(\tau,\beta) = (3/4)\pi\cdot\dot{q}_{Rad}\cdot(\tau + \mu + 2/3) \quad (4.2)$$

where again $\mu = \cos\beta$. If τ increases, the relative importance of μ disappears gradually. For $\tau = 1$ and $\tau = 15$ we have $i'(1,0^0)/i'(1,90^0) = 1.6$ and $i'(15,0^0)/i'(15,90^0) = 1.064$, respectively. Van de Hulst and Grossmann [1968] also calculate an almost isotropic "escape" function i' in a thick atmospheric layer. Calculations made by Kattawar and Plass [1976] for the radiance in optically thick media (Rayleigh scattering in ocean and clouds) show that the angular dependence of the radiance and the polarization is independent of the incident distribution of radiation. Furthermore, Hottel et al. [1971] report no dependence of transmission coefficients on anisotropic phase functions. However, none of these references considered the large anisotropy factors μ that Caps et al. [1984] included in the calculations.

As a consequence, interpretation of absorption lines in planetary atmospheres is made difficult [van de Hulst and Grossmann 1968 p.52].

4.7 Conclusion

The question how large τ_0 must be to allow correct application of the two-flux and diffusion models has been answered using the isotropy criterion for directional intensity and the calculations in

Fig.3.2. The good agreement between Monte Carlo simulation and LAS model found for $\tau_0 \geq 10$ shows in addition that it is sufficient to know τ_0 and ω_1 for a reliable calculation of $\dot{q}_{Rad}$ even if strong anisotropic scattering is present. These findings greatly facilitate the description of anisotropic scattering in nontransparent media.

5. Measurement of Temperature-Dependent Thermal Conductivity and Extinction Coefficient

5.1 Expected Temperature Dependence of Extinction Coefficients of Real Materials

The theoretical temperature dependence of gaseous and solid thermal conductivity λ_{Gas} and λ_S, respectively, has already been reviewed in Sects.2.2,3. It can be concluded from Sect.2.4.5 that a possible temperature dependence of extinction coefficient E is not only due to the known temperature dependence of the local black-body radiation intensity that is used in (3.40) to calculate the Rosseland mean $E_R(T)$ of spectral E_Λ. In contrast, measured absorption coefficients $A_\Lambda(T)$ of Al_2O_3, MgO and other nondispersed solids reported very recently by Cabannes and Billard [1987] show a pronounced *intrinsic* temperature dependence, $A_\Lambda(T) \propto T^2$ or even T^3 (Fig.2.23). If the solids are dispersed, a scattering contribution S_Λ will add to $A_\Lambda(T)$ to give the total $E_\Lambda(T)$. Since S_Λ predominantly relies on the dispersed nature of this solid, it can be expected that S_Λ = const and that the temperature dependence of $E_\Lambda(T)$ is thus weaker than the temperature dependence of $A_\Lambda(T)$. A linear temperature dependence of $E_\Lambda(T)$ on T was observed also for fused silica in the temperature range $0 \le T \le 1800°C$ at a few selected wavelengths ($9.6\ \mu m \le \Lambda \le 10.61\ \mu m$) [McLachlan and Meyer 1987].

The intrinsic temperature dependence of $E_\Lambda(T)$ adds to the theoretical variations of E with temperatures that arise from the $e_{\Lambda b}(T)$ and $e_b(T)$ in (3.40).

This *thermodynamic* temperature dependence of E is subject to the spectral function E_Λ. We neglect, for the moment, an intrinsic dependence of E_Λ on T. [If, in addition, the medium is grey, we have $E_R(T) \equiv E_\Lambda$ from (3.40)]. Figure 5.1 shows a schematic classification of

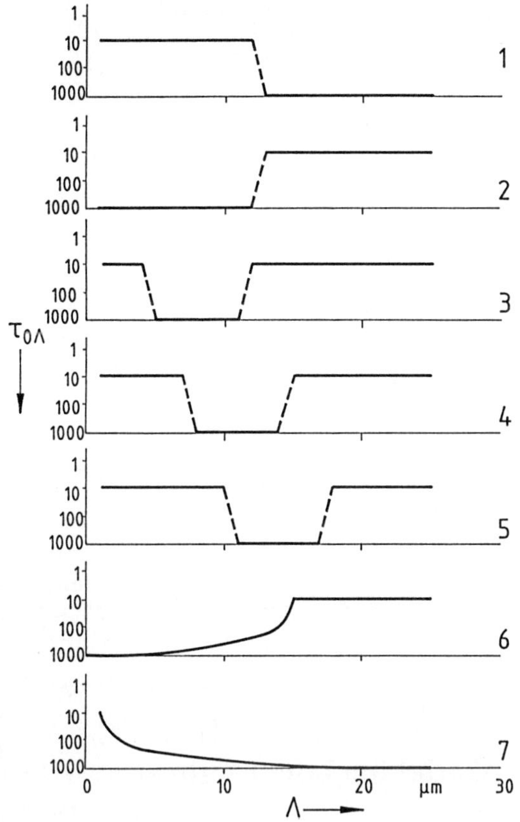

Fig.5.1. A rough separation of extinction (i.e. optical thickness) spectra of inorganic substances into seven categories that correspond approximately to the spectral extinction properties of TiO_2 (1), WB_5 (2), SiO_2 or $Na_2B_4O_7$ (3-5), Fe_3O_4 (6) and KH_2PO_4 (7). For a comparison with real spectra (experimental values) see the collection of [Nyquist and Kagel 1971]

spectral dependencies $\tau_{0,\Lambda}$ (or E_Λ) of nondispersed inorganic substances. The diagrams have been selected according to numerous experimental spectra presented by Nyquist and Kagel [1971]. Although fine structures are completely neglected, it is the "macroscopic" behaviour of the spectra that is responsible for a temperature dependence of $E_R(T)$. Eq.(3.40) was used to calculate $E_R(T)$ in the range 200 K $\leq$ T $\leq$ 2400 K. Fig.5.2 reveals that strongly temperature dependent extinction coefficients can be expected below 900 K for most of the substances. At elevated temperatures, τ_{0R} (or E_R) is almost constant since the maximum of the black body emissive power is located at wavelengths below 3 μm, and the $\tau_{0,\Lambda}$ spectra No.1-6 are constant in this region.

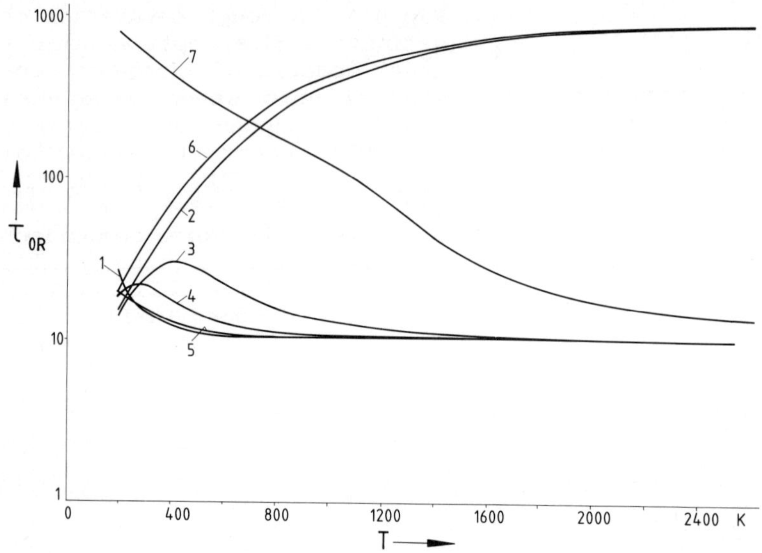

Fig.5.2. Calculated optical thicknesses τ_{OR} (Rosseland mean) using the spectra of Fig.5.1 ($1\ \mu m \leq \Lambda \leq 25\ \mu m$) as a function of temperature

5.2 Predictions of the Diffusion Model for Local Values of Heat Flow Components

Measurements of local thermal conductivities $\lambda(x_i)$ in order to determine temperature dependent solid thermal conductivities $\lambda_S(T)$ and temperature dependent extinction coefficients $E^*(T)$ in evacuated dispersed media have been suggested recently [Reiss and Ziegenbein 1983, 1985a,c]. The experimental method uses a simultaneous measurement of total heat flow through, and temperature profile within, the homogeneous or inhomogeneous medium (Sect.5.3).

Figure 2.10 has already shown the temperature dependence of λ_S of glass fibres (and spheres) taking into account the experimental temperature dependence of $\lambda_{SM}(T)$ and $Y(T)$. In Fig.5.3, $E_R(T)$ of glass fibres has been calculated using (3.40) and an experimental spectrum E_Λ measured between $2\ \mu m \leq \Lambda \leq 50\ \mu m$ at room temperature. From both figures, $\lambda_S(T)$ and $E_R(T)$ of glass fibres can roughly be approximated by a linear relation between $400\ K \leq T \leq 700\ K$. Since it will be shown

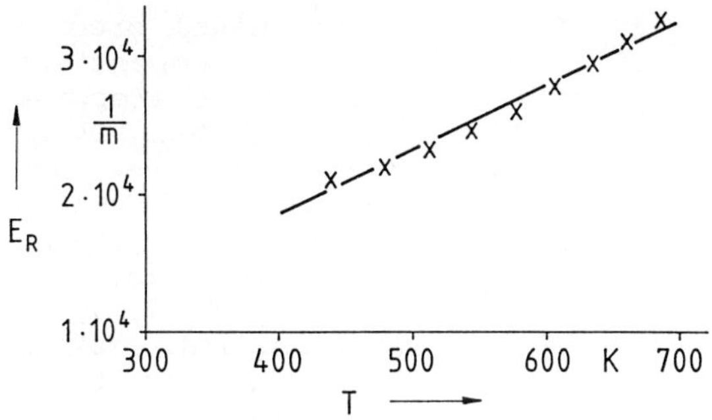

Fig.5.3. Rosseland mean $E_R(T)$ of experimental spectral extinction coefficients of glass fibres ($2\ \mu m \leq \Lambda \leq 50\ \mu m$). From [Reiss and Ziegenbein 1983]

in Sect.5.3 that the curvature of $\lambda(T^{*3})$ diagrams observed for glass fibres can be explained by these simple temperature dependencies, a correction of the extinction coefficient because of an intrinsic (possibly nonlinear) temperature dependence was neglected. As a consequence, E(T) extracted from the curved $\lambda(T^{*3})$ contain both theoretical and possibly existing intrinsic temperature dependencies in a linear approximation.

Linear relations for the temperature dependence of $\lambda_S(T)$ and E(T) (3.49,50) have been used to calculate the temperature profile and local heat flow components $\dot{q}_{Cond}$ and $\dot{q}_{Rad}$ by (3.48) in a homogeneous medium (Fig.5.4, curves labeled with b). For comparison, the corresponding results for temperature independent λ_{Cond} and E are also given (label a). All curves a and b in Fig.5.4 refer to the same τ_0 and the same (thickness averaged, case a) λ_S. Obviously, only very small differences between the temperature profiles a and b result from the calculation. The curvature of profile a is due to the small conduction/radiation parameter N_1. The experimental method suggested above thus depends to a large extent on the accuracy of temperature measurement.

Using the same τ_0 as in Fig.5.4 but a large conduction/radiation parameter N_1, the temperature profile calculated as before for case a is almost exactly linear (Fig.5.5). Note that $\dot{q}_{Rad}$ does *not* vanish.

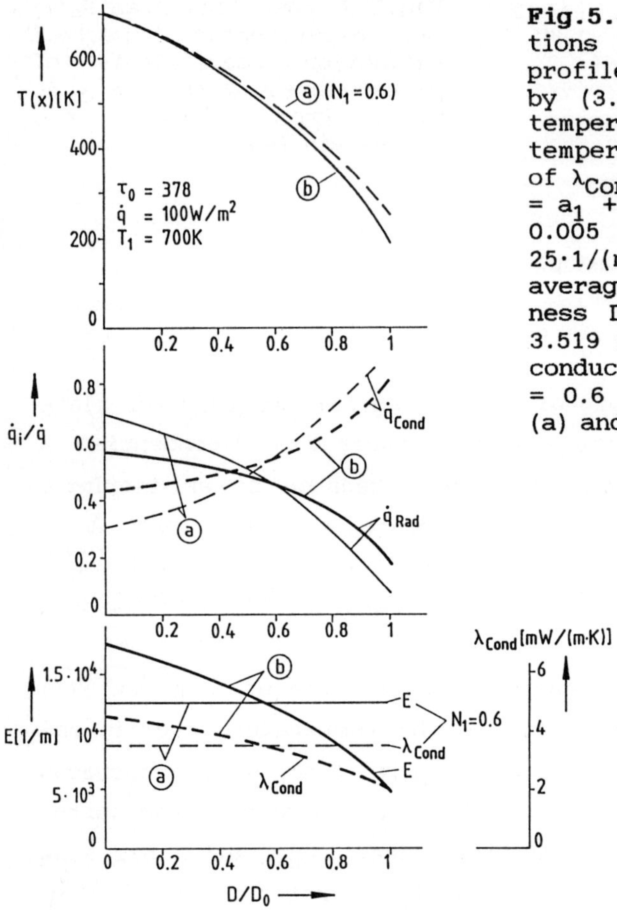

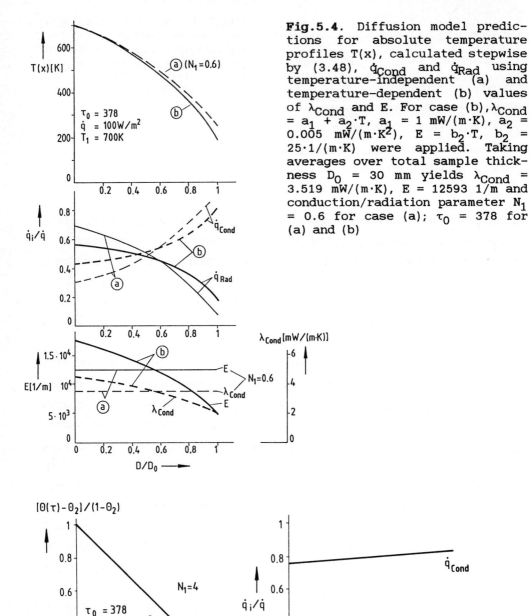

Fig.5.4. Diffusion model predictions for absolute temperature profiles T(x), calculated stepwise by (3.48), $\dot{q}_{Cond}$ and $\dot{q}_{Rad}$ using temperature-independent (a) and temperature-dependent (b) values of λ_{Cond} and E. For case (b), $\lambda_{Cond} = a_1 + a_2 \cdot T$, a_1 = 1 mW/(m·K), a_2 = 0.005 mW/(m·K^2), $E = b_2 \cdot T$, b_2 = 25·1/(m·K) were applied. Taking averages over total sample thickness D_0 = 30 mm yields λ_{Cond} = 3.519 mW/(m·K), E = 12593 1/m and conduction/radiation parameter N_1 = 0.6 for case (a); τ_0 = 378 for (a) and (b)

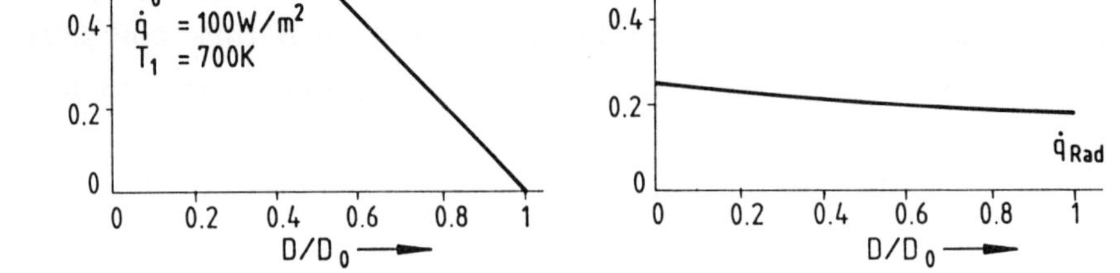

Fig.5.5. Caption see opposite page

5.3 Experimental Procedure and Results

Local thermal conductivities $\lambda(x_i)$ are obtained via

$$\lambda(x_i) = -\dot{q}\cdot\frac{x_i - x_{i+1}}{T(x_i) - T(x_{i+1})} \tag{5.1}$$

where $\dot{q}$ is the total (constant) heat flow and the $T(x_k)$ are measured by temperature sensors distributed in the sample at positions x_k. As before, we can plot local $\lambda(x_i)$ versus T^{*3} using

$$T_{ij}^{*3} = \left[T^2(x_i) + T^2(x_{i+1})\right]\cdot[T(x_i) + T(x_{i+1})] \tag{5.2}$$

according to the definition of T^{*3} given in (3.52). Figure 5.6 is based on the same τ_0 and the same (thickness averaged, case a) λ_{Cond} as in Fig.5.4 for the calculation of $\lambda(T^{*3})$ by (3.48) for cases a and b defined as before. If the temperature sensors are very closely spaced,

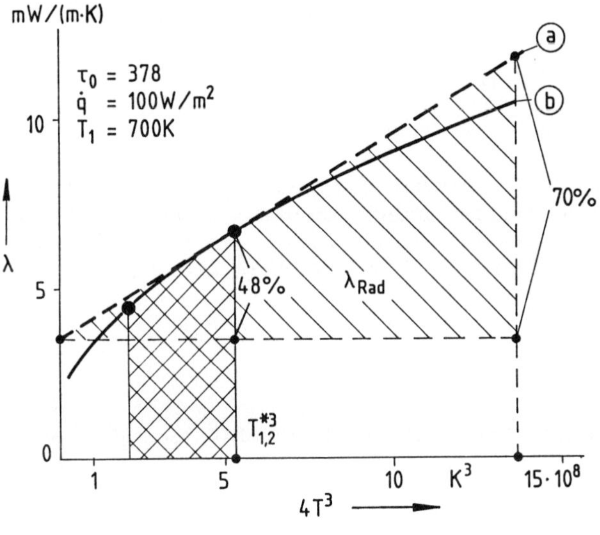

Fig.5.6. Stepwise-calculated diagram $\lambda(T^{*3})$ using (3.48) and the same λ_{Cond} and E (cases a and b) as in Fig.5.4. The shaded region denotes the considerably extended interval on the $4T^3 = T^{*3}$ axis that can be covered if $\lambda(T_{ij}^{*3})$ instead of $\lambda(T_{1,2}^{*3})$ (cross-hatched region) is measured (with $T_i \simeq T_{i+1} = T$, i.e. using closely spaced temperature sensors in the medium)

Fig.5.5. Diffusion model predictions for dimensionless temperature profile, calculated by (3.48), $\dot{q}_{Cond}$ and $\dot{q}_{Rad}$ using the same temperature-independent E and τ_0 as in Fig.5.4 (case a) but a temperature-independent λ_{Cond} = 25 mW/(m·K) that yields N_1 = 4.

$T(x_i) \simeq T(x_{i+1}) = T$, so that $T^{*3} \simeq 4T^3$, which is used as abscissa in Fig.5.6. As follows immediately from (3.51), $\lambda(T^{*3})$ is linear in T^{*3} for case a. Curved $\lambda(T^{*3})$ diagrams have to be expected for temperature dependent $\lambda_{Cond}(T)$ and $E(T)$ (case b). The figure shows, in addition, that $T_{ij}{}^{*3}$ extends to considerably higher values than are achievable with the usual $T_{1,2}{}^{*3}$ defined by wall temperatures T_1 and T_2. Accordingly, this experimental method facilitates time consuming measurements of $\lambda(T_{1,2}{}^{*3})$ and improves the distinction between curved and linear $\lambda(T^{*3})$.

Curved experimentally determined $\lambda(T^{*3})$ diagrams have been obtained from low-temperature experiments (Fig.5.7) and at high temperatures (Fig.5.8). Kaganer's observations (Fig.5.7) were, however, not analyzed in terms of temperature dependent $\lambda_S(T)$ and $E(T)$.

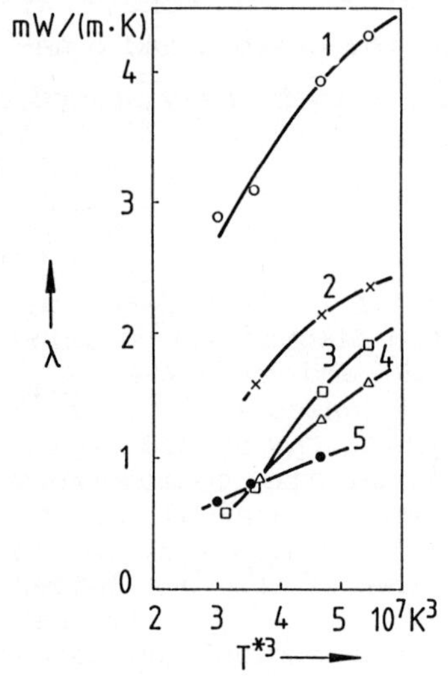

Fig.5.7. Curved diagrams $\lambda(T^{*3})$ observed with silica under vacuum. Labels 1 to 5 correspond to grain diameters d = 1.5, 5.3, 13, 8.5 and 11.7 μm [Kaganer 1969a p.78]

Using (3.42,48-50) for linear temperature dependent $\lambda_{Cond}(T)$ and $E(T)$, we have for total λ

$$\lambda = a_1 + a_2 \cdot T + \frac{16\sigma \cdot n^2}{3b_2} \cdot T^2 . \tag{5.3}$$

This equation transforms into

$$\lambda = C_0 + C_1 \cdot (T^{*3})^{1/3} + C_2 \cdot (T^{*3})^{2/3} \qquad \text{if} \tag{5.4}$$

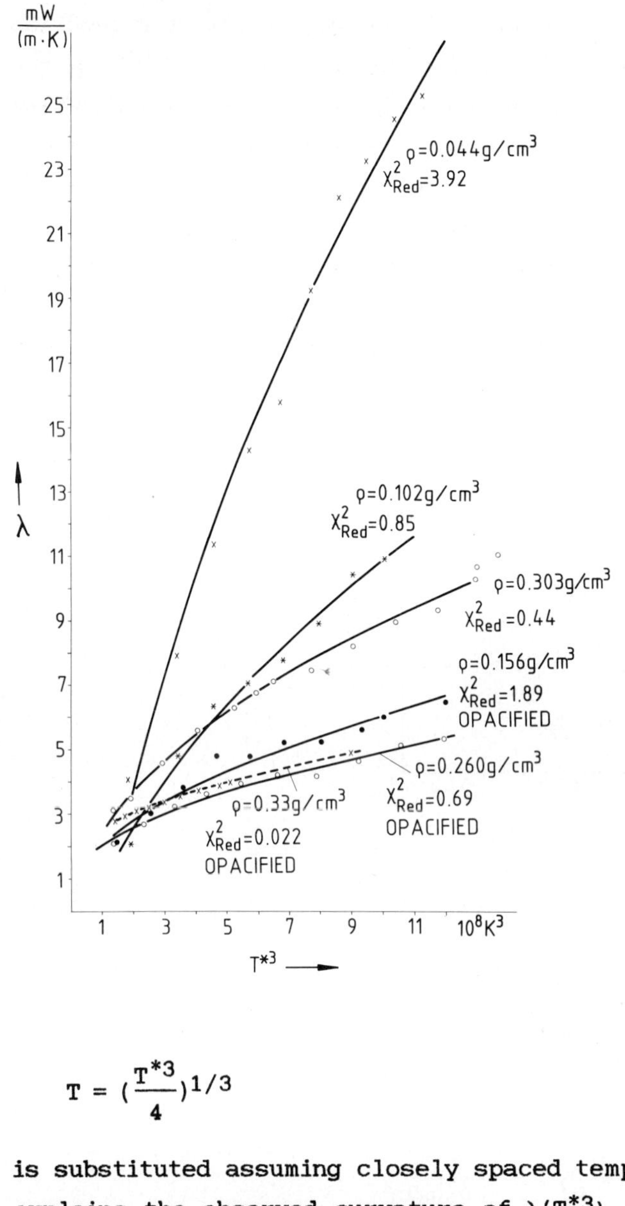

Fig.5.8. Curved diagrams $\lambda(T^{*3})$ of pure and opacified glass fibres of different densities ρ. Symbols denote datapoints, solid and dashed curves are least squares fits to the data which are in addition controlled by χ^2 tests (5% experimental errors assumed). [Reiss and Ziegenbein 1985c], reprinted by permission of Pion Ltd.

$$T = \left(\frac{T^{*3}}{4}\right)^{1/3}$$

is substituted assuming closely spaced temperature sensors. Eq.(5.4) explains the observed curvature of $\lambda(T^{*3})$. A fit of λ using (5.3) to experimental values finally yields the constants a_1, a_2 and b_2.[1]

Since (5.4) is based on well-founded experimental values for $\lambda_S(T)$ and E(T), it follows explicitly that linear diagrams $\lambda(T^{*3})$ for glass fibres cannot exist, at least not in the range of temperatures to

[1] A relation $E(T) = b_1 + b_2 \cdot T$ $(b_1 \neq 0)$ does not yield a closed form solution for a_1, a_2, b_1 and b_2.

which the linear approximations for $\lambda_S(T)$ and $E(T)$ apply. The question arises whether other apparently linear relations for $\lambda(T^{*3})$ in the literature are due to scattering of data that do not allow a decision between linear or curved $\lambda(T^{*3})$ or due to the experimental T^{*3} intervals, being too small to allow a decision.

Temperature-dependent $\lambda_S(T)$, $E^*(T)$ and $E^*(T)/\rho$ for pure and heavily doped glass fibres of different density ρ are shown in Fig.5.9

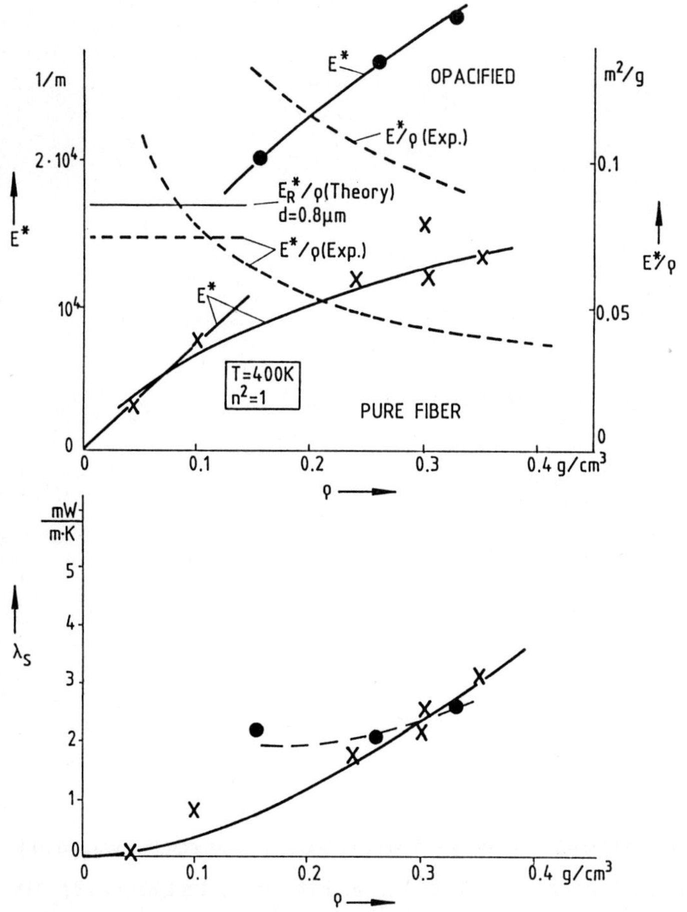

Fig.5.9. Experimental E^* and λ_S values of pure and opacified glass fibres extracted at T = 400 K from the curved $\lambda(T^{*3})$ diagrams in Fig.5.8. The data (symbols) are given in dependence of density ρ. At low densities ($\rho \leq 150$ kg/m^3), E^* was approximated by a linear relation $E^* \propto \rho$, at high densities a curved dependence $E(\rho)$ yields better agreement with data; note the good agreement with Mie theory (thin horizontal line) if independent scattering is assumed (i.e. E^*/ρ = const) at low densities (d denotes particle diameter). From [Reiss and Ziegenbein 1985c] reprinted by permission of Pion Ltd.

[Reiss and Ziegenbein 1985c]. At low densities ($\rho \leq 100$ kg/m^3), the experimental E^*/ρ of pure glass fibres are in good agreement with theory (thin horizontal line) if independent scattering is assumed, i.e. if E^* increases linearly with density. At higher densities, the observed E^* are apparently no longer a linear function of density but considerably smaller. Does this observation indicate the onset of dependent scattering in a fibrous medium?

5.4 Can Calorimetric Measurements of Extinction Coefficients Reveal Dependent Scattering?

An estimate of the parameter $(D-d)/\Lambda$ (Sect.2.4.3) using the expression $\delta = 0.785d/(1-\Pi)$, derived by Verschoor and Greebler [1952] for the pore diameter δ in aggregates of fibres, yields the following. If 50 kg/m$^3 \leq \rho \leq 350$ kg/m^3, $\rho_0 = 2610$ kg/m^3, $d = 0.8$ µm and $(D-d) = \delta$ we then have 33 µm $\geq (D-d) \geq 5$ µm. This gives $(D-d)/\Lambda \leq 0.3$ for $\Lambda \geq 30$ µm and $\rho \geq 175$ kg/m^3. Another rough estimate for the effect of dependent scattering on Q_{Ext} is made by application of an expression given by Hottel et al. [1971]

$$\log\left[\log(Q_{Ext}/Q_{Ext}^{Eff})\right] = 0.25 - 5.1(D-d)/\Lambda \ . \tag{5.5}$$

This yields $\rho \geq 100$ kg/m^3 as a critical density with respect to wavelengths $\Lambda \geq 10$ µm. Both estimates are in rough agreement with the observation (Fig.5.9) that a deviation of E^* becomes significant at $\rho \geq 100$ kg/m^3. Brewster's and Hottel's relations are, however, reported for spherical particles. It is not at all clear that these relations are also applicable to cylindrical particles. Caps [1985] investigated experimentally the reduction of specific extinction coefficients of glass fibres under external pressure load p between $0 \leq p \leq 90$ bar. The extinction coefficient taken at $\Lambda = 6$ µm was reduced by a factor of about two only when full load was applied. Full load corresponds to a porosity of about 0.55, i.e. a density $\rho \geq 1600$ kg/m^3. This density is considerably larger than $\rho \leq 400$ kg/m^3

where the curved E^* in Fig.5.9 deviate by a factor of two from the linearly extrapolated E^*. Obviously, dependent scattering, if present at all, cannot account fully for the curvature of $E^*(\rho)$ seen in Fig.5.9. The E^* reported by Grunert et al. [1969a] for glass fibres (d = 0.6 μm) also follow a linear relationship with ρ for $\rho \leq 260$ kg/m^3.

5.5 Inhomogeneous Media; Outlook for a Completion of the Method

An application of the experimental method for a determination of the temperature dependence of $\lambda_S(T)$ and $E(T)$ in inhomogeneous media is reported by Reiss and Ziegenbein [1985c].

In agreement with a prediction of the diffusion model [Reiss and Ziegenbein 1985c] and theoretical work [Gogol 1979] there is only a slight change in total heat flow if the arrangement of two half-layers of pure and heavily doped glass fibres with respect to the heat source and sink is reversed.

Although previous investigations [Wakao and Kato 1969] included a measurement of temperature profiles and local thermal conductivities, an analysis in terms of temperature dependent $\lambda_S(T)$ and $E(T)$ apparently has not been presented prior to this work.

Naturally, an extension of the method to include gaseous thermal conductivity at low temperatures according to

$$\lambda = a_1 + a_2 \cdot T + a_3 \cdot T^2 + a_4 \cdot T^{1/2} \tag{5.6}$$

would be desirable because this would finally complete the method from its present state (separation of $\lambda_{Cond} = \lambda_S + \lambda_{Gas}$) to a determination of a second isolated component (λ_{Gas} in addition to λ_{Rad}).

6. Optimum Radiation Extinction

6.1 Formulation of the Complete Optimization Concept

The observed resonances in Q_{Ext} of spherical dielectric particles (Fig.2.15) have been used in the literature to define optimum particle diameters d_{Opt} for the creation of optimum radiation extinction coefficients E_Λ at a given wavelength Λ. However, Figs.4.1,17 have demonstrated the importance of anisotropic scattering by spherical and cylindrical particles. It cannot be expected that the d_{Opt} for maximum Q_{Ext}/x (or E) coincide with the d_{Opt} for maximum E^*. As a consequence, an optimization of E [Kaganer 1969a p.41] that is useful only in the case of *isotropic* scattering is not sufficient with respect to the scattering properties of real substances: though preferentially treated in the literature, isotropic scattering is not the normal but rather the *exceptional* case in the whole catalogue of scattering phenomena.

Besides anisotropic scattering, inclined incidence has to be accounted for if the particles are cylindrical. This is important if particle axes cannot by external means be oriented perpendicular to temperature gradient. For example, using a glass fibre paper as shown in Fig.4.5, such an (at least preferential) orientation is easily performed. However, investigation of scattering properties of ice needles or of glass wool has to account for a statistical distribution of orientations with respect to incoming radiation.[1]

[1] In a quiet atmosphere, an orientation of hexagonal ice needles can occur that is responsible for the halo scattering phenomenon. Also polarization of radiation can be caused by interstellar dust particles that are oriented parallel to each other by interstellar magnetic fields.

The scattering problem of irregularly oriented cylinders with diameter d and length ℓ both of which are large compared with wavelength has been treated by elementary diffraction theory [van de Hulst 1981 pp.109,110]. The position of bright and dark rings is slightly shifted if the diameters of large spherical or irregularly oriented cylindrical particles are identical.

For arbitrary d and ℓ, the effective extinction coefficient for cylindrical particles

$$E_\Lambda{}^*(\phi) = \frac{4\rho}{\pi\cdot\rho_0}\cdot\frac{Q_{Ext,\Lambda}(\phi)}{d}\cdot[1 - \Omega_\Lambda\cdot\mu_\Lambda(\phi)] = \frac{4\rho}{\pi\cdot\rho_0}\cdot\frac{Q^*_{Ext,\Lambda}(\phi)}{d} \tag{6.1}$$

with respect to an angle of incidence ϕ has to be averaged over all ϕ

$$E_\Lambda{}^* = \int_0^{\pi/2} E_\Lambda{}^*(\phi)\cdot\cos\phi\cdot d\phi\ . \tag{6.2}$$

However, instead of calculating the explicit average (6.2) [Wang and Tien 1983] it has been shown that a reliable calculation of averaged $Q^*_{Ext,\Lambda}(\phi)$ from the corresponding extinction cross sections at normal incidence is possible [Caps et al. 1985], for instance for $\Omega_\Lambda = 1$,

$$Q^*_{Ext,\Lambda}(\phi) = C_1\cdot(1-\mu_\Lambda)\cdot Q_{Ext,\Lambda}(\phi{=}0) \tag{6.3}$$

where $C_1 = 2/3$. Very good agreement is reported for scattering parameters $x \geq 5$ (whereas the calculations of van de Hulst [1981 p.109] are applicable to about $x \geq 60$ only) and, for a purely scattering medium, if the refractive index n exceeds 2 [Baumeister 1986].

In addition, Caps [1985] has shown that a similar simplification is possible if the fibres are arranged in planes perpendicular to incident ratiation with their axes oriented randomly in the planes.

If the wavelength is fixed, the optimization procedure for the creation of an optimum extinction coefficient of spherical and cylindrical particles is thus reduced to merely a variation of particle diameter. Complete optimization must take into account, however, a wavelength distribution at a given radiation temperature.

6.2 Optimum Particle Diameters for Spheres and Cylinders

Figure 2.16 allows the determination of optimum spherical particle diameters d_{opt} for creating optimum spectral extinction coefficients E_Λ/ρ (this figure summarizes the previous optimization concept which was valid only for isotropic scattering). If anisotropic scattering and a wavelength distribution is accounted for, Fig.6.1a-d show $E_R{}^*/\rho$ for spherical particles calculated from the rigorous Mie theory as a function of particle diameter d and temperature T. Constant refractive indices have been assumed in the calculations. The diagrams demonstrate the existence of an optimum particle diameter in all cases considered. Comparison of Fig.6.1b and d shows that a large real part n of the complex refractive index $m = n - i \cdot k$ is favourable for an increase of $E_R{}^*/\rho$. According to Fig.6.1a and c, a large imaginary part k increases $E_R{}^*/\rho$ even more strongly at elevated temperatures (note that all diagrams have been calculated with constant ρ_0). The optimum particle diameters of the nonmetallic spherical particles are between 2 and 6 μm.

The existence of optimum particle diameters of dielectric cylindrical particles is demonstrated in Fig.6.2 [Caps et al. 1985]; the calculations are performed for perpendicular incidence.

The previous figures demonstrate that large real parts of the refractive index are favourable for exciting optimum radiation extinction by nonmetallic spherical and cylindrical particles. Metallic spherical particles have even larger $E_R{}^*/\rho$ at high temperatures. The influence of large imaginary parts k of m of cylindrical particles will be investigated in the next section.

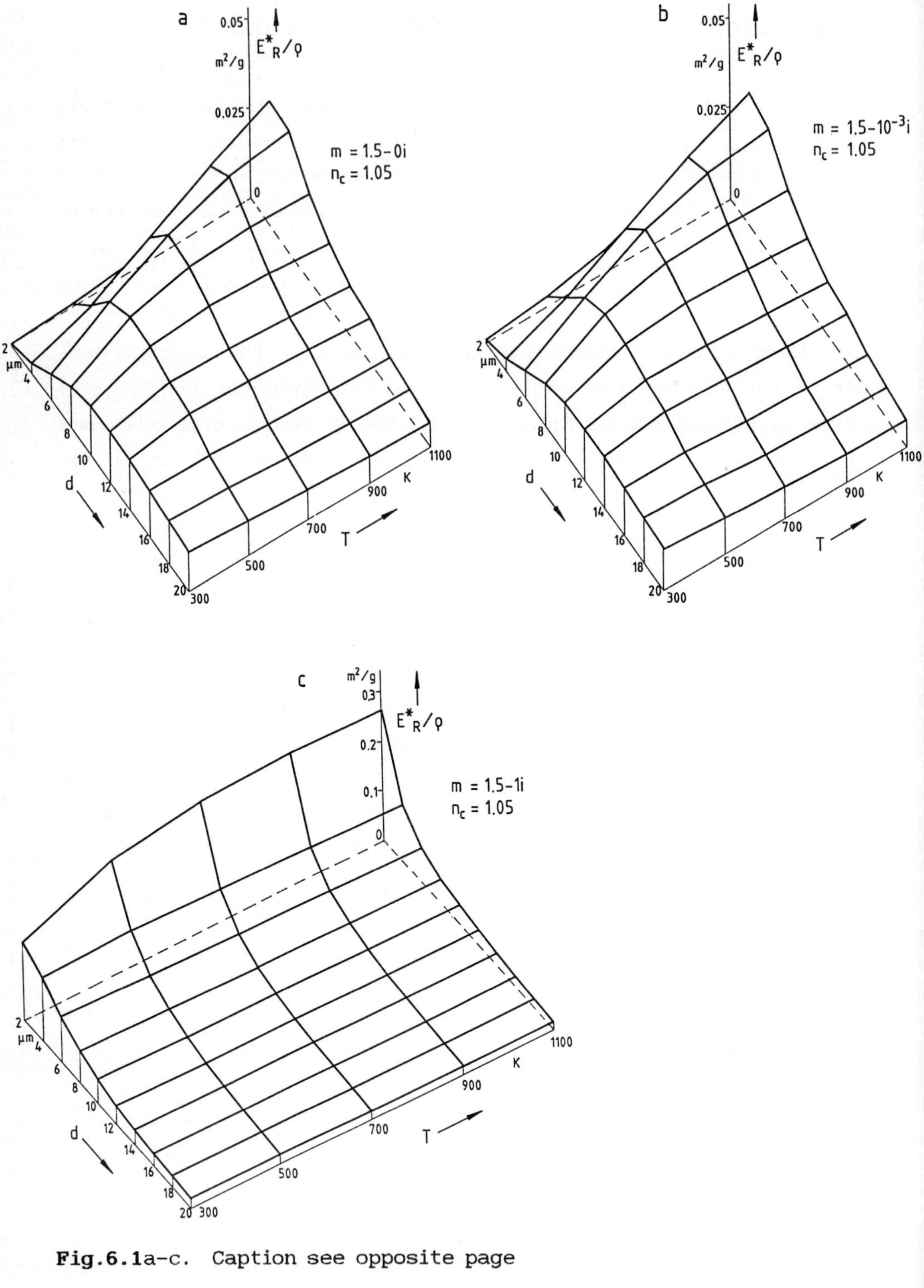

Fig.6.1a-c. Caption see opposite page

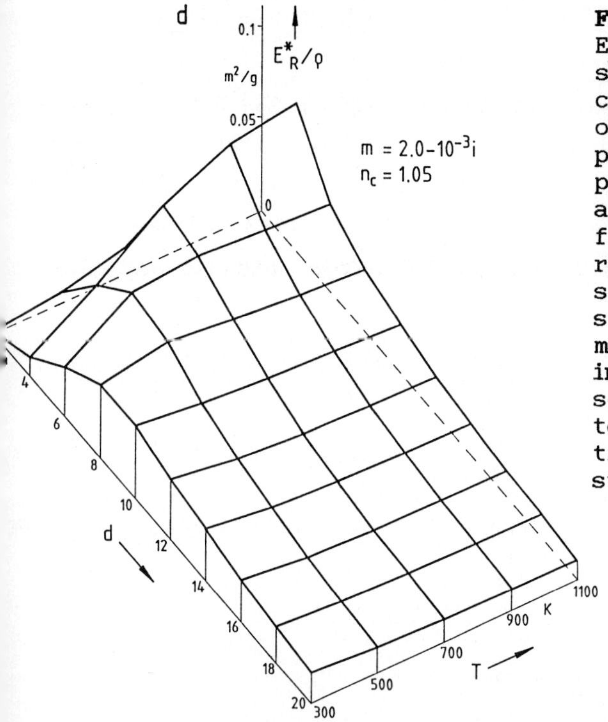

Fig.6.1a-d. Rosseland mean E_R^*/ρ of effective values of specific extinction coefficients calculated from rigorous Mie theory for spherical particles as a function of particle diameter d and radiation temperature T for different complex indices of refraction (ρ denotes density). The diagrams demonstrate the existence of optimum particle diameters also in the case of anisotropic scattering and with respect to a complete thermal spectrum. All diagrams use constant ρ_0

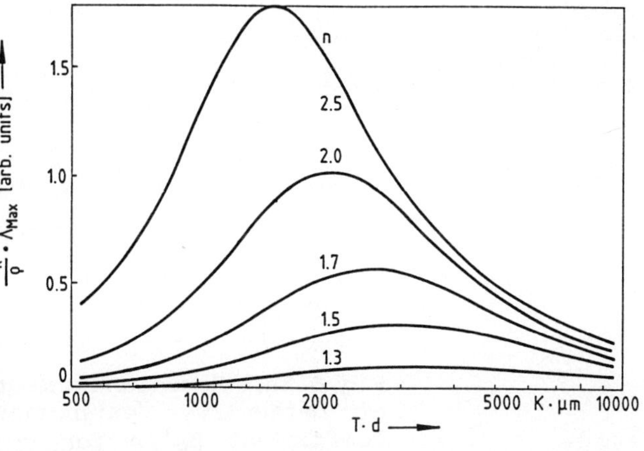

Fig.6.2. Optimum fibre diameters taking into account anisotropic scattering and a complete thermal spectrum calculated from Mie theory for different real indices of refraction [Caps et al. 1985]. The figure is calculated assuming temperature T fixed and particle diameter d variable; λ_{Max} denotes wavelength where maximum of intensity of thermal spectrum occurs at a given T. Reprinted by permission of Pion Ltd.

6.3 Thin Metallic Fibres

It has been known since the work of Kerker [1969 p.289 Fig.6.13] that very large relative extinction cross sections Q_{Ext} can be achieved using thin totally reflecting cylinders. This interesting aspect has been reinvestigated only recently [Wang and Tien 1983; Trunzer 1983; Reiss et al. 1985]. Figure 6.3a,b show $E_R{}^*/\rho$ for metallic and glass

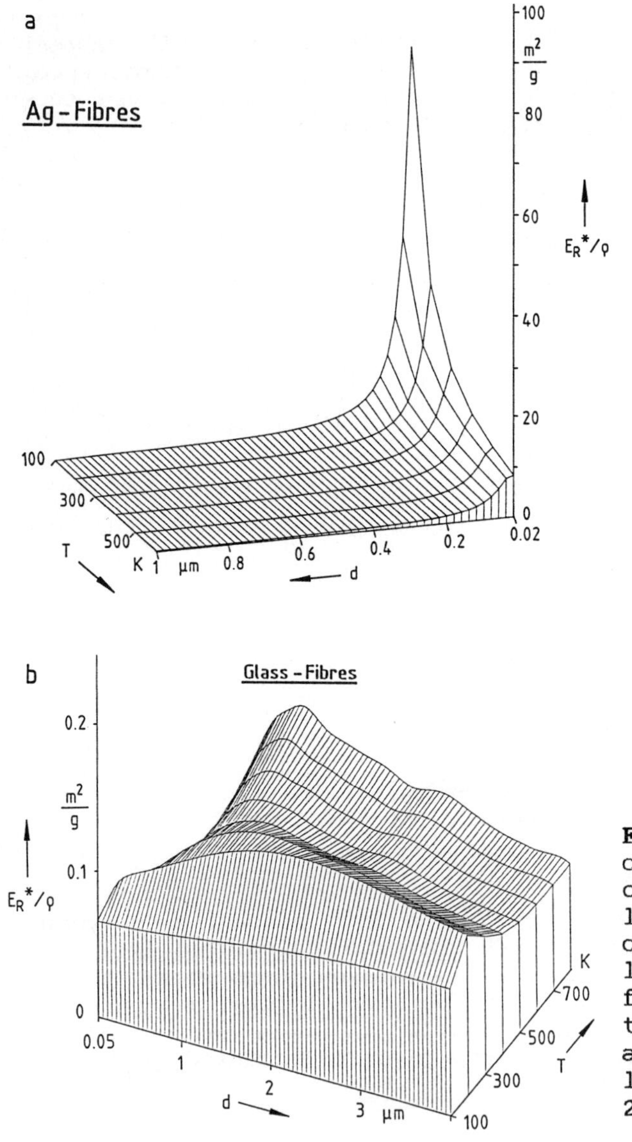

Fig.6.3a,b.Rosseland mean of effective extinction coefficient $E_R{}^*/\rho$ for cylindrical Ag particles (a) or glass fibres (b) calculated from Mie theory as a function of particle diameter d and radiation temperature T within the wavelength intervals 1 $\mu m \leq \Lambda \leq$ 200 μm

fibres calculated from rigorous Mie theory as a function of particle diameter d and temperature T [Reiss et al. 1985]. At low temperatures, extinction by metallic fibres can be larger by a factor of 500 than is obtainable with glass fibres, provided the particle diameter is below 0.1 μm.

Polarization effects do not introduce alterations to the calculated E_R^*/ρ [Reiss 1985a]. However, dependent scattering could require considerable corrections if porosity is not very large (because of the huge Q_{Ext}).

When the experimental investigations were started, thin metallic fibres with diameters below 1 μm were not yet available. Therefore attempts have been made to aluminize conventional glass fibres in a CVD process [Reiss et al. 1985]. Calorimetric measurements show that λ_S of the glass fibres was hardly increased by the metallization. A calculated spectrum was confirmed by a spectroscopic measurement. Although the fibre diameter was still about 4 μm, the E_R^*/ρ of the metallized fibres (about 0.2 m^2/g) was increased by a factor of two compared with the pure fibres. This is already considerably larger than can be achieved with glass fibres that were heavily doped using strongly absorbing powder (Fig.5.9). Further improvements were achieved when glass fibre paper with a fibre diameter of 0.8 μm was sputtered with Ag, Pd or Mo. Extinction coefficients up to 0.45 m^2/g have been measured.

7. Conclusion

Calorimetric, spectroscopic and numerical investigations of the mechanisms of heat transfer in nontransparent media show that radiative transfer can successfully be described as a conduction process in these media. Most of the reviewed work considers the true conductive component of total thermal conductivity and the extinction coefficient as being independent of temperature. However, theoretical estimates of the temperature dependence of gaseous conductivity and solid contact conductivity and of wavelength-averaged extinction coefficients as well as measured temperature profiles show that this is only a rough approximation to the real situation.

As a consequence, the frequently used "additive approximation" seems to be wrong in many cases because the influence of $\lambda_{Rad}(T)$ on the temperature profiles can be very strong. Only if temperature profiles are to a good approximation linear is it allowed to calculate thermal conductivity components independently of each other.

Anisotropic scattering can be included into analysis of radiative transfer in nontransparent media by scaling extinction coefficients to effective values. Spectroscopic measurements of extinction coefficients and numerical calculations using the rigorous Mie theory are in agreement with effective extinction coefficients extracted from calorimetric experiments.

Optimum particle diameters for creation of maximum effective extinction coefficients exist for spherical and cylindrical particles also if a total spectrum of infrared radiation is considered.

Isotropic intensity distributions which are essential for applicability of two-flux approximations are characteristic for nontransparent media even if they possess strongly anisotropic scattering properties. It is not possible to deduce the degree of anisotropic scattering from escape functions in these media.

Prediction of transmission and reflection coefficients by three- and five-flux approximation are in close agreement with Monte-Carlo methods and LAS models.

References

Ambarzumjan V.A., Mustel E.R. et al.: 1957, Theoretische Astrophysik, transl. into German by I. Heller and E.A. Gussmann, in Hochschulbücher für Physik, ed. by F.X. Eder and R. Rompe (VEB Deutscher Verlag der Wissenschaften, Berlin, GDR).

Ammar M.M., Gharib S., Halawa, M.M., El Badry Kh., Ghoneim N.A., El Batal H.A.: 1982, Thermal conductivity of some silicate glasses in relation to composition and structure, J. Non-Cryst. Sol. **53**, pp.165-172.

Amrhein E.: 1974, Absorption der Gläser im fernen IR (10-500 cm^{-1} bzw. 1000-20 μm), in Nachordnungsfelder in Gläsern, Fachausschussbericht No.70 (Deutsche Glastechnische Gesellschaft e.V., Frankfurt) pp.127-140.

Anderson E.E., Viskanta R.: 1973, Spectral and boundary effects on coupled conduction-radiation heat transfer through semitransparent solids, Wärme- und Stoffübertragung **1**, 14-24.

Anderson E.E., Viskanta R., Stevenson W.H.: 1973, Heat transfer through semitransparent solids, Trans. ASME, J. Heat Transfer **5/73**, 179-186.

Ardenne M.v.: 1973, Tabellen zur Angewandten Physik, Vol.III (VEB Deutscher Verlag der Wissenschaften, Berlin, GDR) p.173.

Azad F.H., Modest M.F.: 1981, Evaluation of the radiative heat flux in absorbing, emitting and linear-anisotropically scattering cylindrical media, Trans. ASME, J. Heat Transfer **103**, 350-356.

Bauer W., Steinhardt R.: 1982, Spektrale und Gesamtemissionsgrade von ausgewählten feuerfesten Oxiden und Siliziumkarbid, Silikattechnik **33**, 212-213.

Baule B.: 1914, Theoretische Behandlung der Erscheinungen in verdünnten Gasen, Ann. Physik **44**, 145-176.

Baumeister K.: 1986, Bestimmung der Strahlungsleitfähigkeit disperser Medien durch IR-optische Messungen, Diplom-Thesis, Phys. Inst. der Univ. Würzburg.

Bentsen L.D., Thomas J.R., Hasselmann D.P.H.: 1984, Role of view factor in the effective thermal conductivity/diffusivity of transparent materials, Commun. Am. Ceram. Soc. **5/84**, C90-C91.

Bergmann T.L., Houf W.G., Incropera F.P.: 1983, Effect of single scatter phase function distribution on radiative transfer in absorbing-scattering liquids, Int. J. Heat Mass Transfer **26**, 101-107.

Bett K.E., Saville G.: 1965, The effective thermal conductivity of powdered insulating materials at high gas pressures and temperatures, A.I.Ch.E-I.Chem.E. Symposium Series **2**, 71-76 (London: Inst. Chem. Engrs.).

Bjurström H., Karawacki E., Carlsson B.: 1984, Thermal conductivity of a microporous particulate medium: moist silica gel, Int. J. Heat Mass Transfer **27**, 2025-2036.

Blodgett K.B., Langmuir I.: 1932, Accommodation coefficient of hydrogen; a sensitive detector of surface films, Phys. Rev. **40**, 78-104.

Boccara A.C., Fournier, D.: 1987, "Heat diffusion and fractals", in Photoacoustic and Photothermal Phenomena, ed. by P. Hess and J. Pelzl, Proc. 5th Int'l. Topical Meeting on Photoacoustic and Photothermal Phenomena, Heidelberg, Springer Ser. Opt. Sci., Vol.58 (Springer, Berlin, Heidelberg 1988)

Bohren C.F., Huffman D.R.: 1983, Absorption and Scattering of Light by Small Particles (Wiley, New York).

Born M., Wolf E.: 1965, Principles of Optics, 3rd ed. (Pergamon, Oxford).

Bourrely C., Chiappetta P., Torresani B.: 1986, Light scattering by particles of arbitrary shape: a fractal approach, J. Opt. Soc. Am. **A3**, 250-255.

Braun R., Fischer S., Schaber A.: 1983, Elimination of the radiant component of measured liquid thermal conductivities, Wärme- und Stoffübertragung **17**, 121-124.

Brewster M.Q.: 1981, Radiative transfer in packed and fluidized beds, PhD Thesis, Univ. of California, Berkeley.

Brewster M.Q., Tien C.L.: 1982a, Examination of the two-flux model for radiative transfer in particular systems (shorter communication), Int. J. Heat Mass Transfer **25**, 1905-1907.

Brewster M.Q., Tien C.L.: 1982b, Radiative Transfer in packed fluidized beds: dependent versus independent scattering, Trans. ASME, J. Heat Transfer **104**, 573-579.

Bridgman P.W.: 1923, The thermal conductivity of liquids, Proc. Nat'l. Acad. Sci. USA **9**, 341-345.

Buchenau U., Müller I.: 1972, Optical properties of magnetite, Solid State Commun. **11**, 1291-1293.

Büttner D., Fricke J., Krapf R., Reiss H.: 1983, Measurement of the thermal conductivity of evacuated load-bearing, high-temperature powder and glass board insulations with a 700x700 mm^2 guarded hot-plate device, Proc. 8th Eur. Conf. Thermophys. Prop., Baden-Baden 1982, High Temp.- High Pressures **15**, 233-240.

Büttner D., Fricke J., Reiss H.: 1985a, Thermal conductivity of evacuated load-bearing powder and fibre insulation under variable external load, Proc. 9th Eur. Conf. Thermophys. Prop., Manchester 1984, High Temp.- High Pressures **17**, 333-341.

Büttner D., Fricke J., Reiss H.: 1985b, Analysis of radiative and solid conduction components of the total thermal conductivity of an evacuated glass fiber insulation - Measurements with a 700x700 mm^2 variable load guarded hot plate device, AIAA 20th Thermophys. Conf., Williamsburg 1985, AIAA Conf. Paper No.85-1019.

Cabannes F., Maurau J.-C., Hyrien M., Klarsfeld S.M.: 1979, Radiative heat transfer in fibreglass insulating materials as related to their optical properties, High Temp.- High Pressures **11**, 429-434.

Cabannes F.: 1980a, Les isolants thermiques modernes, Rev. Gén. Therm. Fr. **219**, 183-192.

Cabannes F.: 1980b, Propriétés infrarouges et conductivité thermique par rayonnement des isolants thermiques à fibres réfractaires, Rev. Int. Hautes Tempér. Réfract. Fr. **17**, 120-133.

Cabannes F., Billard D.: 1987, Measurement of infrared absorption of some oxides in connection with the radiative transfer in porous and fibrous materials, Int. J. Thermophys. **8**, 97-118.

Caps R., Fricke J., Reiss H.: 1983a, Improving the extinction properties of an evacuated high-temperature powder insulation, Proc. 8th Eur. Conf. Thermophys. Prop., Baden-Baden 1982, High Temp.- High Pressures **15**, 225-232.

Caps R., Trunzer A., Büttner D., Fricke J., Reiss H.: 1983b, Spectral transmission and reflection properties of high temperature insulation materials and their relation to radiative heat flow, in Thermal Conductivity 18, Proc. 18th Int. Thermal Conductivity Conf., Rapid City 1983, ed. by T. Ashworth and D.R. Smith (Plenum, New York 1985) pp.403-411.

Caps R., Fricke J.: 1983, Radiative heat transfer in highly transparent silica aerogel, Phys. Inst. der Univ. Würzburg, Rep. E12-1283-1.

Caps R., Trunzer A., Büttner D., Fricke J., Reiss H.: 1984, Spectral transmission and reflection properties of high temperature insulation materials, Int. J. Heat Mass Transfer **27**, 1865-1872.

Caps R., Fricke J., Reiss H.: 1985, Radiative heat transfer in anisotropically scattering fiber insulations, Proc. 9th Eur. Conf. Thermophys. Prop., Manchester 1984, High Temp.- High Pressures **17**, 303-309.

Caps R.: 1985, Strahlungswärmeströme in evakuierten thermischen Superisolationen, Doktor-Thesis, Phys. Inst. der Univ. Würzburg.

Caren R.P.: 1969, Radiation transfer from a metal to a finely divided particulate medium, Trans. ASME, J. Heat Transfer **2/69**, 154-156.

Chan C.K., Tien C.L.: 1974a, Radiative transfer in packed spheres, Trans. ASME, J. Heat Transfer **2/74**, 52-58.

Chan C.K., Tien C.L.: 1974b, Combined radiation and conduction in packed spheres, in Heat Transfer 1974, Proc. 5th Int. Heat Transfer Conf., Tokyo 1974, Vol.I, (The Japanese Soc. of Mech. Eng., The Japanese Soc. of Chem. Eng., Japan) pp.72-74.

Chandrasekhar S.: 1960, Radiative Transfer (Dover, New York).

Chapman S.: 1912, The kinetic theory of a gas constituted of spherically symmetrical molecules, Philos. Trans. (A)**211**, 433-483.

Chupp R.E., Viskanta R.: 1974, Development and evaluation of a remote sensing technique for determining the temperature distribution in semitransparent solids, Trans. ASME, J. Heat Transfer **8/74**, 391-397.

Cockett A.H., Molnar W.: 1960, Recent improvements in insulants, Cryogenics **9/60**, 21-26.

Coenen M.: 1974, Wärmeleitung in Gläsern, in Nahordnungsfelder in Gläsern, Fachausschussbericht No.70 (Deutsche Glastechnische Gesellschaft e.V., Frankfurt) pp.43-60.

Cunnington G.R., Tien C.L.: 1972, Heat transfer in microsphere cryogenic insulation, Proc. Cryog. Eng. Conf., Boulder 1972, Paper No.C-1, Adv. Cryog. Eng. **18**, 103-111.

Dayan A., Tien C.L.: 1975, Heat transfer in a gray planar medium with linear anisotropic scattering, J. Heat Transfer **8/75**, 391-396.

Defay R., Prigogine I., Bellemans A., Everett D.H.: 1966, Surface Tension and Adsorption (Longmans Green, London).

Deimling A.: 1984, Wärmedämmung im Vakuum-Stahlmantelrohrsystem (Berichtsanteil), Bundesministerium für Forschung und Technologie, Rep. No. T84-167.

Deissler R.G., Eian C.S.: 1952, Investigation of effective thermal conductivities of powders, NACA Research Memorandum NACA-RM-E 52-C05.

Deresiewicz H.: 1958, Mechanics of granular matter, in Advances in Applied Mechanics, Vol.5, ed. by H.L. Dryden, Th. von Karman and G. Kuerti (Academic, New York).

Dietz P.W.: 1979, Effective thermal conductivity of packed beds, Ind. Eng. Chem. Fundam. **18**, 283-286.

Domoto G.A., Wang W.C.: 1974, Radiative transfer in homogeneous non-gray gases with nonisotropic particle scattering, Trans. ASME, J. Heat Transfer **8/74**, 385-390.

Drolen B.L., Kumar S., Tien C.L.: 1987, Experiments on dependent scattering of radiation, AIAA 22nd Thermophys. Conf., Honolulu 1987, AIAA Conf. Paper No. 87-1485.

Drolen B.L., Tien C.L.: 1987, Independent and dependent scattering in packed-sphere systems, J. Thermophys. Heat Transfer **1**, 63-68.

Dushman S.: 1958, Scientific Foundations of Vacuum Technique, 4th pr. (Wiley, New York).

Espe W.: 1960, Werkstoffkunde der Hochvakuumtechnik, Vol.II (VEB Deutscher Verlag der Wissenschaften, Berlin, GDR).

Eucken A.: 1913, Über das Wärmeleitvermögen, die spezifische Wärme und die innere Reibung der Gase, Phys. Z. **XIV**, 324-332.

Eucken A.: 1940, Allgemeine Gesetzmässigkeiten über das Wärmeleitvermögen verschiedener Stoffarten und Aggregatzustände, Forsch. Geb. Ingenieurwes. **11**, 6-20.

Fahrenkrog H.: 1983, FG Techn. Glas, Wertheim, FRG, private communication.

Favro D.L., Shepard S.M., Kuo P.K., Thomas R.L.: 1987, Mechanism for the generation and scattering of sound and thermal waves in thermoacoustic microscopes,in Photoacoustic and Photothermal Phenomena, ed. by P.Hess and J. Pelzl, Proc. 5th Int'l. Topical Meeting on Photoacoustic and Photothermal Phenomena, Heidelberg, Springer Ser. Opt. Sci., Vol.58 (Springer, Berlin, Heidelberg 1988).

Fernandes R., Francis J., Reddy J.N.: 1980, A finite element approach to combined conductive and radiative heat transfer in a planar medium, AIAA 15th Thermophys. Conf., Snowmass 1980, AIAA Conf. Paper No. 80-1487; Heat transfer and thermal control, AIAA Thermophys. Prog. in Aeronautics and Astronautics **78**, 92-109.

Fernandes R., Francis J.: 1982, Combined conductive and radiative heat transfer in an absorbing, emitting and scattering cylindrical medium, Trans. ASME, J. Heat Transfer **104**, 594-601.

Fine H.A., Jury S.H., Yarbrough D.W., McElroy D.L.: 1980, Analysis of heat transfer in building insulation, Oak Ridge Natl. Lab., Oak Ridge, Rep. ORNL/TM-7481.

Fried E.: 1969, Thermal conduction contribution to heat transfer at contacts, in Thermal Conductivity, Vol.2, ed. by R.P. Tye (Academic, London).

Gaskell P.H., Johnson D.W.: 1976, The optical constants of quartz, vitreous silica and neutron-irradiated vitreous silica (I), J. Non-Cryst. Solids **20**, 153-169.

Glasstone S. Edlund M.C.: 1961, Kernreaktortheorie, transl. into German by W. Glaser and H. Grümmer (Springer, Vienna).

Glicksman L., Schuetz M., Sinofsky M.: 1987, Radiation heat transfer in foam insulations, Int. J. Heat Mass Transfer **30**, 187-197.

Gogel W., 1979: An effect of asymmetry of the rate of the heat flow and of the effective thermal conductivity, Proc. 6th Eur. Conf. Thermophys. Prop., Dubrovnik 1978, High Temp.- High Pressures **11**, 491-498.

Goldsmid H.J., Kaila M.M.: 1980, Heat flow at pressed contacts, Aust. J. Phys. **33**, 745-752.

Goldsmith A., Waterman Th.E., Hirschhorn H.J.: 1961, Handbook of Thermophysical Properties of Solid Materials, Vol.III Ceramics (Pergamon, Oxford).

Grunert W.E., Notaro F., Reid R.L.: 1969a, Opacified fibrous insulations, AIAA 4th Thermophys. Conf., San Francisco 1969, AIAA Conf. Paper No.69-605.

Grunert W.E., Mohihara H., Reid R.L., Massing P.N.: 1969b, Guarded flat plate thermal conductivity apparatus for testing multi-foil insulations in the 20°C - 1000°C range, Proc. 9th Intern. Thermal Cond. Conf., Iowa State Univ., Ames 1969, USAEC, CONF-691002, Physics (TID-4500), pp.658-672.

Hamaker H.C.: 1947, Radiation and heat conduction in light-scattering material II: General equations including heat conduction, Philips Res. Rep. **2**, 103-111.

Hamann C.: 1983, Dielektrische und ferroelektrische Erscheinungen in Festkörpern, in Atom- und Kernphysik, 1st ed.(Harry Deutsch, Thun, Switzerland) p.563.

Harris L.: 1955, Preparation and infrared properties of aluminum oxide films, J. Opt. Soc. Am. **45**, 27-29.

Harris L., Piper J.: 1962, Transmittance and reflectance of aluminum oxide films in the far infrared, J. Opt. Soc. Am. **52**, 223-224.

Heaslett M.A., Warming R.F.: 1965, Radiative transport and wall temperature slip in an absorbing planar medium, Int. J. Heat Mass Transfer **8**, 979-994.

Hellwege K.H.: 1981, Einführung in die Festkörperphysik, 2nd ed. (Springer, Berlin, Heidelberg).

Hertz H.: 1882, Über die Berührung fester elastischer Körper, J. Reine Angew. Math. **29**, 156-171.

Holm R.: 1967, Electric Contacts (Springer, Berlin, Heidelberg).

Hottel H.C.: 1962, Radiation as a diffusion process, Int. J. Heat Mass Transfer **5**, 82-83.

Hottel H.C., Sarofim A.F.: 1967, Radiative Transfer (McGraw-Hill, New York).

Hottel H.C., Sarofim A.F., Vasalos I.A., Dalzell W.H.: 1970, Multiple scatter: Comparison of theory with experiment, Trans. ASME, J. Heat Transfer **5/70**, 285-291.

Hottel H.C., Sarofim A.F., Dalzell W.H., Vasalos I.A.: 1971, Optical properties of coatings. Effect of pigment concentration, AIAA J. **9/71**, 1895-1898.

Hsieh C.K., Su K.C.: 1979, Thermal radiative properties of glass from 0.32 to 206 μm, Sol. Energy **22**, 37-43.

Jackson J.D.: 1967, Classical Electrodynamics, 6th pr. (Wiley, New York).

Jackson K.W., Black W.Z.: 1983, A unit cell model for predicting the thermal conductivity of a granular medium containing an adhesive binder, Int. J. Heat Mass Transfer **26**, 87-99.

Jaenicke W.: 1956, Lichtstreuung und Aufhellungsvermögen weißer Pigmente, Z. Elektrochem. **60**, 163-174.

Jakob M.: 1964, Heat Transfer, Vol.I, 9th pr. (Wiley, New York).

Kaganer M.G., Semenova R.S.: 1967, An investigation of the thermal diffusivity and thermal conductivity of insulating powders at atmospheric pressure and in vacuum by various methods, J. Eng. Phys. (USSR) **12/79**, 15-18, Engl. transl. of Inzh.-Fiz. Zh. **13**, 24-30.

Kaganer M.G.: 1969a, Thermal Insulation in Cryogenic Engineering, transl. by A. Moscona, Israel Progr. Sci. Transl., Jerusalem, Engl. transl. of Teplovaya Izolyatsiya v Tekhnike Nizkikh Temperatur (Izdatel'stvo "Mashinostroenie", Moscow 1966).

Kaganer M.G.: 1969b, Untersuchung der Ausbreitung von Licht im streuenden Medium durch die Methode der Diskreten Ordinaten, Opt. Spektrosk. **XXVI**, 443-449, transl. into German by G. Wahl, Heidelberg, 1981.

Kattawar G.W., Plass G.N.: 1976, Asymptotic radiance and polarization in optically thick media: ocean and clouds, Appl. Opt. **15**, 3166-3178.

Kattawar G.W., Dean C.E.: 1983, Electromagnetic scattering from two dielectric spheres: comparison between theory and experiment, Opt. Lett. **8**, 48-50.

Keesom W.H., Schmidt G.: 1936, Researches of heat conduction by rarefied gases. II: The thermal accommodation coefficient of helium, neon, hydrogen and nitrogen on glass at 70-90 K, Physica **III**, 1085-1092.

Kennard E.H.: 1938, Kinetic Theory of Gases with an Introduction to Statistical Mechanics (McGraw-Hill, London).

Kerker M.: 1969, The Scattering of Light and Other Electromagnetic Radiation (Academic, London).

Kingery W.D., Klein J.D., McQuarrie M.C.: 1958, Development of ceramic insulating materials for high temperature use, Trans. ASME **80**, 705-710.

Klemens P.G.: 1955, The scattering of low-frequency lattice waves by static imperfections, Proc. Phys. Soc. A **LXVIII**, 1113-1128.

Klemens P.G.: 1969, Theory of thermal conductivity of solids, in: Thermal Conductivity, Vol.1, ed. by R.P. Tye (Academic, London) pp.1-68.

Klemens P.G.: 1983, Theory of thermal conductivity of solids at high temperatures, Proc. 8th Eur. Conf. Thermophys. Prop., Baden-Baden 1982, High Temp.- High Pressures **15**, 249-254.

Klemens P.G.: 1985, Radiative heat transfer in composites, Proc. 9th Eur. Conf. Thermophys. Prop., Manchester 1984, High Temp.- High Pressures **17**, 381-385.

Klett D.E., Irey R.K.: 1968, Experimental determination of thermal accommodation coefficients, preprint, Univ. of Florida, Gainesville.

Knudsen M.: 1911, Die Molekulare Wärmeleitung der Gase und der Akkommodationskoeffizient, Ann. Physik **34**, 593-656.

Kortüm G., Oelkrug D.: 1964, Über den Streukoeffizienten der Kubelka-Munk-Theorie, Z. Naturforsch. **19a**, 28-37.

Kostylev V.M.: 1964, The thermal conductivity of dispersed materials at different atmospheric pressures, High Temp. **2**, 15-21.

Kourganoff V.: 1952, Basic Methods in Transfer Problems (Clarendon, Oxford).

Krikorian O.H.: 1985, Estimation of the thermophysical and mechanical properties and the equation of state of Li_2O, Proc. 9th Eur. Thermophys. Conf. Thermophys. Prop., Manchester 1984, High Temp.- High Pressures **17**, 161-172.

Kruse P.W., McGlauchlin L.D., McQuistan R.B.: 1971, Grundlagen der Infrarottechnik (Verlag Berliner Union, Stuttgart), German Transl. of Elements of Infrared Technology (Wiley, New York) by P. Becker.

Kunc T., Lallemand M., Saulnier J.B.: 1984, Some new developments on coupled radiative-conductive heat transfer in glasses - experiments and modelling, Int. J. Heat Mass Transfer **27**, 2307-2319.

Landau, L.D., Lifschitz E.M.: 1965, Lehrbuch der theoretischen Physik, Vol.VII Elastizitätstheorie (Akademie Verlag, Berlin, GDR).

Landolt-Börnstein: 1962, Zahlenwerte und Funktionen, 6th ed., Vol.II Eigenschaften der Materie in ihren Aggregatzuständen, Part 8: Optische Konstanten (Springer, Berlin, Göttingen, Heidelberg).

Landolt-Börnstein: 1972, Zahlenwerte und Funktionen, 6th ed., Vol.IV Eigenschaften der Materie in ihren Aggregatzuständen, Part 4b: Wärmetechnik (Springer, Berlin, Heidelberg).

Lang M.L., Wolfe W.L.:1983, Optical constants of fused silica and sapphire from 0.3 to 25 μm, Appl. Opt. **22**, 1267-1268.

Larkin B.K., Churchill St.W.: 1959, Heat transfer by radiation through porous insulations, A.I.Ch.E.J. **5**, 467-474.

Laubitz M.J.: 1959, Thermal conductivity of powders, Can. J. Phys. **37**, 798-808.

Le Doussal H., Bisson G.: 1980, La mesure de la conductivité thermique des produits réfractaires, Rev. Gén. Therm. Fr. **219**, 223-230.

Lee H., Buckius R.O.: 1983, Reducing scattering to nonscattering problems in radiation heat transfer, Int. J. Heat Mass Transfer **26**, 1055-1062.

Leibfried G., Schlömann E.: 1954, Wärmeleitung in elektrisch isolierenden Kristallen, Nachr. Adak. Wiss. Göttingen, Math.-Phys. Kl. **4**, 71-93.

Liou K.N., Cai Q., Barber P.W., Hill S.C.: 1983, Scattering phase matrix comparison for randomly hexagonal cylinders and spheroids, Appl. Opt. **22**, 1684-1687.

Little R.C., Carpenter F.G., Deitz V.R.: 1962, Heat transfer in intensively outgassed powders, J. Chem. Phys. **37**, 1896-1898.

Loeb L.B.: 1961, The Kinetic Theory of Gases, 3rd ed. (Dover, New York).

Lowan A.N., Morse P.M., Feshbach H., Lax M.: 1946, Scattering and Radiation from Circular Cylinders and Spheres, Tables of Amplitudes and Phase Angles, U.S. Navy, Washington, D.C.

Luikov A.V.: 1966, Heat and Mass Transfer in Capillary-porous Bodies, Engl. transl. P.W.B. Harrison, ed. by W.M. Pun (Pergamon, Oxford).

Luikov A.V., Shashkov A.G., Vasiliev L.L., Fraiman Yu.E.: 1968, Thermal conductivity of porous systems, Int. J. Heat Mass Transfer **11**, 117-140.

Madhusudana C.V., Fletcher L.S.: 1981, Gas conductance contribution to contact heat transfer, AIAA 16th Thermophys. Conf., Palo Alto 1981 AIAA Conf. Paper No.81-1163.

Maheu B., Letoulouzan J.N., Gousbet G.: 1984, Four-flux models to solve the scattering transfer equation in terms of Lorenz-Mie parameters, Appl. Opt. **23**, 3353-3362.

Mandelbrot B.B.: 1983, The Fractal Geometry of Nature (Freeman, New York).

Maxwell-Garnett J.C.: 1904, Colours in metal glasses and in metallic films, Philos. Trans. R. Soc. London, Ser. A, Vol.CCIII, pp.385-420.

McElroy D.L., Moore J.P.: 1969, Radial heat flow methods for the measurement of the thermal conductivity of solids, in: Thermal Conductivity Vol.1, ed. by R.P. Tye (Academic, London) pp.185-239.

McKay N.L., Timusk T.: 1984, Determination of optical properties of fibrous thermal insulation, J. Appl. Phys. **55**, 4064-4071.

McKellar B.H.J., Box M.A.: 1981, The scaling group of the radiative transfer equation, J. Atmos. Sci. **38**, 1063-1068.

McLachlan A.D., Meyer F.P.: 1987, Temperature dependence of the extinction coefficient of fused silica for CO_2 laser wavelengths, Appl. Opt. **26**, 1728-1731.

Meissner H.P., Michaels A.S., Kaiser R.: 1964, Crushing strength of zinc oxide agglomerates, I&EC Process Design and Dev. **3**, 202-205.

Michels W.C.: 1932, Accommodation coefficients of helium and argon against tungsten, (Letter to the Editor) Phys. Rev. **40**, 472-273.

Mie G.: 1908, Beiträge zur Optik trüber Medien, speziell kolloidaler Metallösungen, Ann. Physik **25**, 377-445.

Mihalas D.: 1967, The calculation of model stellar atmospheres, in Methods in Computational Physics, Vol.7: Astrophysics, ed. by B. Alder, S. Fernbach and M. Rotenberg (Academic, New York).

Mil'man S.B., Kaganer M.G.: 1975, Study of radiative transfer in vacuum-powder insulation by infrared spectroscopy (Plenum, New York) Engl. transl. from Inzh.-Fiz. Zh. **28**, 40-45.

Mulder C.A.M., van Lierop J.G.: 1986, Preparation, densification and characterization of autoclave dried SiO_2 gels, in Aerogels, ed. by J. Fricke, 1st Int. Symp., Würzburg 1985, Springer Proc. Phys., Vol.6 (Springer, Berlin, Heidelberg) pp.68-75.

Neuroth N.: 1974, Aussagen der Spektroskopie im nahen und mittleren Infrarot zur Glasstruktur, in Nahordnungsfelder in Gläsern, Fachausschussbericht Nr.70 (Deutsche Glastechnische Gesellschaft e.V., Frankfurt) pp.141-187.

Nowobilski J.J.: 1979, Insulation development for high temperature batteries for electric vehicle application; final report prepared for

Department of Energy by Union Carbide, Linde Division, Tonawanda, Contr. No.EM-78-C-01-5160.

Nyquist R.A., Kagel R.O.: 1971, Infrared Spectra of Inorganic Compounds (3800 to 45 cm^{-1}) (Academic, London).

Ordal M.A., Long L.L., Bell R.J., Bell S.E., Bell R.R., Alexander Jr. R.W, Ward C.A.: 1983, Optical properties of the metals Al, Co, Cu, Au, Fe, Pb, Ni, Pd, Pt, Ag, Ti and W in the infrared and far infrared, Appl. Opt. **22**, 1099-1119.

Pelanne C.M.: 1979, Does the insulation have a thermal conductivity? The revised ASTM Test Standards require an answer, Authorized reprint from J. Special Techn. Publ. 660, ASTM, Philadelphia.

Pelanne C.M.: 1981, Discussion on experiments to separate the "effect of thickness" from systematic equipment errors in thermal transmission measurements, Authorized reprint from J. Special Techn. Publ. 718, ASTM, Philadelphia.

Percus J.K., Yevick G.J.: 1958, Analysis of classical statistical mechanics by means of collective coordinates, Phys. Rev. **110**, 1-13.

Plass G.N.: 1964, Mie scattering and absorption cross sections for aluminum oxide and magnesium oxide, Appl. Opt. **3**, 867-872.

Raines B.: 1939, The accommodation coefficient of helium on nickel, Phys. Rev. **56**, 691-695.

Rapier A.C., Jones T.M., McIntosh, J.E.: 1963, The thermal conductance of uranium dioxide/stainless steel interfaces, Int. J. Heat Mass Transfer **6**, 397-416.

Reddy J.N., Murty V.D.: 1978, Finite-element solution of integral equations arising in radiative heat transfer and laminar boundary-layer theory, Numer. Heat Transfer **1**, 389-401.

Reiss H.: 1981a, An evacuated, load-bearing powder insulation for a high temperature Na/S battery, in: Thermal Conductivity 17, Proc. 17th Int. Thermal Conductivity Conf., Gaithersburg 1981, ed. by J.G. Hust (Plenum, London 1983) pp.569-586.

Reiss H.: 1981b, An evacuated powder insulation for a high temperature Na/S battery, AIAA 16th Thermophys. Conf., Palo Alto 1981, AIAA Conf. Paper No.81-1107.

Reiss H.: 1983, Evacuated, load-bearing powder insulation for high temperature applications, J. Energy **7**, 152-159.

Reiss H., Ziegenbein B.: 1983, Temperature dependent extinction coefficients and solid thermal conductivities of glass fiber insulations, in Thermal Conductivity 18, Proc. 18th Int. Thermal Conductivity Conf., Rapid City 1983, ed. by T. Ashworth and D.R. Smith (Plenum, London 1985) pp.413-424.

Reiss H., Ziegenbein B.: 1985a, An experimental method to determine temperature-dependent extinction coefficients and solid thermal conductivities of glass fibre insulations in calorimetric measurements, Int. J. Heat Mass Transfer **28**, 459-466.

Reiss H., Ziegenbein B.: 1985b, Can thermal conductivity, λ, and extinction coefficient, E, be measured simultaneously? (Technical Note), Int. J. Heat Mass Transfer **28**, 1408-1411.

Reiss H., Ziegenbein B.: 1985c, Analysis of the local thermal conductivity in inhomogeneous glass fiber insulations, Proc. 9th Eur. Conf. Thermophys. Prop., Manchester 1984, High Temp.- High Pressures **17**, 403-412.

Reiss H., Schmaderer F., Wahl G., Ziegenbein B., Caps R.: 1985, Experimental investigation of extinction properties and thermal conductivity of metal-coated dielectric fibers in vacuum, AIAA 20th Thermophys. Conf., Williamsburg 1985, AIAA Conf. Paper No.85-1020, and Int. J. Thermophys. (1987) **8**, 263-280.

Reiss H.: 1985a, Strahlungstransport in dispersen nichttransparenten Medien, Habilitationsschrift, Phys. Inst. der Univ. Würzburg.

Reiss H.: 1985b, Wärmeströme in Isolationen, Spektrum der Wissenschaft **11/85**, 112-125.

Ritzow G.: 1934, Die Temperaturstrahlung glühender Oxyde und Oxydgemische im ultraroten Spektralgebiet, Ann. Physik **19**, 769-799.

Roberts J.K.: 1930, The exchange of energy between gas atoms and solid surfaces, Proc. R. Soc. London **129**, 146-161.

Roberts J.K.: 1933, The exchange of energy between gas atoms and solid surfaces, III. - The accommodation coefficient of neon, Proc. R. Soc. London **142**, 518-524.

Rohatschek H.: 1976, Zur praktischen Durchführung der Wärmeleitfähigkeitsmessung mit der Kugelsonde, Int. J. Heat Mass Transfer **19**, 1433-1439.

Rosseland S.: 1931, Astrophysik auf atomtheoretischer Grundlage, in Struktur der Materie in Einzeldarstellungen, ed. by M. Born and J. Franck (Verlag von Julius Springer, Berlin).

Roux, J.A., Smith A.M.: 1981, Determination of radiative properties from transport theory and experimental data, AIAA 16th Thermophys. Conf., Palo Alto 1981, AIAA Conf. Paper No.81-1168.

Rumpf H.: 1961, The strength of granules and agglomerates, in Agglomeration, Int. Symposium, PA 1961, ed. by W.A. Knepper, (Interscience, London) pp.379-418.

Russell H.W.: 1935, Principles of heat flow in porous insulators, J. Am. Ceram. Soc. **18**, 1-5.

Saxena S.C., Afshar R.: 1985, Thermal accommodation coefficient of gases on controlled solid surfaces: argon-tungsten system, Int. J. Thermophys. **6**, 143-163.

Scheuerpflug P., Caps R., Büttner D., Fricke J.: 1985, Apparent thermal conductivity of evacuated SiO_2-aerogel tiles under variation of radiative boundary conditions, Int. J. Heat Mass Transfer **28**, 2299-2306.

Schlegel A., Alvarado S.F., Wachter P.: 1979, Optical properties of magnetite (Fe_3O_4), J. Phys. C **12**, 1157-1164.

Schulze G.E.R.:1967, Metallphysik (Akademie Verlag, Berlin, GDR).

Schuster A.: 1905, Radiation through a foggy atmosphere, Astrophys. J. **XXI**, 1-22.

Schwarzschild K.: 1906, Über das Gleichgewicht der Sonnenatmosphäre, Nachr. Akad. Wiss. Göttingen, pp.41-53.

Serebryanyi G.L., Zarudnyi L.B., Shorin S.N.: 1968, Measurement of the heat conductivity coefficient of vacuum-powder insulation at high temperatures (Plenum, New York) Engl. transl. from Teplofiz. Vys. Temp. **6**, 547-548.

Siegel R., Howell J.R.: 1972, Thermal Radiation Heat Transfer (McGraw-Hill Kogakusha, Tokyo).

Slack G.A.: 1973, Nonmetallic crystals with high thermal conductivity, J. Phys. Chem. Solids **34**, 321-335.

Sparrow E.M., Cess R.D.: 1966, Radiation Heat Transfer, Thermal Science Series, ed. by R.G. Eckert (Brooks/Cole, Belmont).

Springer G.S.: 1971, Heat transfer in rarefied gases, in Advances in Heat Transfer, Vol.7, ed. by Th.F. Irvine jr. and J.P. Hartnett (Academic, New York) pp.163-218.

Stephens G.L.: 1984, Scattering of plane waves by soft obstacles: anomalous diffraction theory for circular cylinders, Appl. Opt. **23**, 954-959.

Stratton J.A.: 1941, Electromagnetic Theory (McGraw-Hill, New York).

Strong H.M., Bundy F.P., Bovenkerk H.P.: 1960, Flat panal vacuum thermal insulation, J. Appl. Phys. **31**, 39-50.

Sutherland W.: 1893, The viscosity of gases and molecular force, Philos. Mag. **36**, 507-531.

Thomas L.B.: 1967, Thermal accommodation of gases on solids, in Fundamentals of Gas-Surface Interactions, ed. by H. Saltsburg, J.N. Smith Jr. and M. Rogers (Academic, London) pp.346-369

Tong T.W., Tien C.L.: 1980, Analytical models for thermal radiation in fibrous insulations, J. Therm. Insulation **4**, 27-44.

Tong T.W., Tien C.L.: 1983, Radiative heat transfer in fibrous insulations - Part I: Analytical study, Trans. ASME, J. Heat Transfer **105**, 70-75.

Touloukian Y.S., ed.: 1970, Thermophysical Properties of Matter, Vol.7 Thermal Radiative Properties - Metallic Elements and Alloys, Vol.8 Thermal Radiative Properties - Nonmetallic Solids (IFI/Plenum Data Corp., New York).

Truelove J.S.: 1984, The two-flux model for radiative transfer with strong anisotropic scattering (Technical Note), Int. J. Heat Mass Transfer **27**, 464-466.

Trunzer A.: 1983, Untersuchung der spektralen Extinktion im Infrarotbereich von Glasfasern und Trübungsmitteln zur Minimierung der thermischen Strahlungsleitfähigkeit in Hochtemperaturisolationen, Zulassungsarbeit zur wissenschaftl. Prüfung für das Lehramt an Gymnasien in Bayern, Phys. Inst. der Univ. Würzburg.

Tsederberg N.V.: 1965, Thermal Conductivity of Gases and Liquids, transl. by Scripta Technica, ed. by R.D. Cess (MIT Press, Cambridge).

Twersky, V.: 1975, Transparency of pair-correlated, random distributions of small scatterers, with applications to the cornea, J. Opt. Soc. Am. **65**, 524-530.

Tye R.P., Desjarlais A.O.: 1983, Factors influencing the thermal performance of thermal insulations for industrial applications; thermal insulation, materials, and systems for energy conservation in the '80s, ASTM STP 789, ed. by F.A. Govan, D.M. Greason and J.D. McAllister (Am. Soc. for Testing and Materials, Philadelphia) pp.733-748.

Unsöld A.: 1968, Physik der Sternatmosphären, corrected reprint of 2nd ed. (Springer, Berlin, Heidelberg).

van de Hulst H.C., Grossmann K.: 1968, Multiple light scattering in planetary atmospheres, in The Atmospheres of Venus and Mars, ed. by J.C. Brandt and M.B. McElroy (Gordon and Breach, New York) pp.35-55.

van de Hulst H.C.: 1980, Multiple Light Scattering Tables, Formulas and Applications, Vol.2 (Academic, New York).

van de Hulst H.C.: 1981, Light Scattering by Small Particles (Dover, New York) corrected printing of 1957 edition (Wiley, New York).

van der Held E.F.M.: 1952, The contribution of radiation to the conduction of heat, Appl. Sci. Res. **A3**, 237-249.

Veale C.R.: 1972, Fine Powders, Preparation, Properties and Uses (Appl. Sci. Publ., London).

Verschoor J.D., Greebler P.: 1952, Heat transfer by gas conduction and radiation in fibrous insulations, Trans. ASME, J. Heat Transfer **8/74**, 961-968.

Viskanta R., Grosh R.J.: 1962a, Effect of surface emissivity on heat transfer by simultaneous conduction and radiation, Int. J. Heat Mass Transfer **5**, 729-734.

Viskanta R., Grosh R.J.: 1962b, Heat transfer by simultaneous conduction and radiation in an absorbing medium, Trans. ASME, J. Heat Transfer **2/62**, 63-72.

Viskanta R., Grosh R.J.: 1964, Heat transfer in a thermal radiation absorbing and scattering medium, Int. Devel. in Heat Transfer, Part 4, Trans. ASME, New York, pp.820-828.

Viskanta R.: 1965, Heat transfer by conduction and radiation in absorbing and scattering materials, Trans. ASME, J. Heat Transfer **2/65**, 143-150.

Viskanta R.: 1966, Radiation transfer and interaction of convection with radiation heat transfer, in Advances in Heat Transfer, Vol.3, ed. by Th.F. Irvine Jr. and J.P. Hartnett (Academic, London) pp.175-251.

Viskanta R.: 1982, Radiation heat transfer: interaction with conduction and convection and approximate methods in radiation, in Heat Transfer 1982, Proc. 7th Int. Heat Transfer Conf., Munich 1982, Vol.1, ed. by U. Grigull, E. Hahne, K. Stephan and J. Straub (Hemisphere, Washington, D.C.) pp.103-121.

Volz F.E.: 1983, Infrared specular reflectance of pressed crystal powders and mixtures, Appl. Opt.**22**, 1842-1855.

Vortmeyer D.: 1979, Wärmestrahlung in dispersen Feststoffsystemen, Chem.-Ing.-Tech. **51**, 839-851.

Wakao N., Kato K.: 1969, Effective thermal conductivity of packed beds, J. Chem. Eng. Jpn. **2**, 24-33.

Wakao N., Vortmeyer D.: 1971, Pressure dependency of effective thermal conductivity of packed beds, Chem. Eng. Sci. **26**, 1753-1765.

Walther A., Dörr J., Eller E.: 1953, Mathematische Berechnung der Temperaturverteilung in der Glasschmelze mit Berücksichtigung von Wärmeleitung und Wärmestrahlung, Glastechnische Berichte **5/53**, 133-140.

Wang R.T., Greenberg J.M., Schuermann D.W.: 1981, Experimental results of dependent light scattering by two spheres, Opt. Lett. **6**, 543-545.

Wang K.Y., Tien C.L.: 1983, Radiative heat transfer through opacified fibers and powders, J. Quant. Spectrosc. Radiat. Transfer **30**, 213-223.

Waxler R.M., Cleek G.W.: 1973, The effect of temperature and pressure on the refractive index of some oxide glasses, J. Res. NBS, Sect. A **77A**, 755-763.

Weaver J.H., Krafka C., Lynch D.W., Koch E.E.: 1981, Optical properties of metals, Part I: The transition metals, in Physics Data, ed. by H. Behrens and G. Ebel (Fachinformationszentrum Energie, Physik, Mathematik GmbH, Karlsruhe,FRG).

Weber H.H.: 1957, Deckfähigkeitsmessungen an Weißpigmenten, Farbe und Lack, 63.Jahrg. **12**, 586-594.

Weber H.H.: 1960, Über das optische Verhalten von kugeligen, isotropen Teilchen in verschiedenen Medien, Kolloid-Z. **174**, 66-72.

Wertheim M.S.: 1963, Exact solution of the Percus-Yevick integral equation for hard spheres, Phys. Rev. Lett. **10**, 321-323.

Wickramasinghe N.C.: 1972, Light Scattering Functions for Small Particles with Applications in Astronomy (Adam Hilger, Bristol, UK).

Wiesmann H.J.: 1983, Energietransport in ebenen Schichten durch gekoppelte Wärmestrahlung und Festkörperwärmeleitung, Rep. KLR 83-182B (Brown Boveri & Cie AG, Baden, Switzerland).

Williams M.M.R.: 1984, The effect of anisotropic scattering on the radiant heat flux through an aerosol, J. Phys. D **17**, 1617-1630.

Wiscombe W.J.: 1977, The delta-M method: rapid yet accurate radiative flux calculations for strongly asymmetric phase functions, J. Atmos. Sci. **34**, 1408-1422.

Woodside W.: 1958, Calculation of the thermal conductivity of porous media, Can. J. Phys. **36**, 815-823.

Wray J.H., Neu J.T.: 1969, Refractive index of several glasses as a function of wavelength and temperature, J. Opt. Soc. Am. **59**, 774-776

Wu S.T., Ferguson R.E., Altgilbers L.L.: 1980, Application of finite-element techniques to the interaction of conduction and radiation in a participating medium, AIAA 15th Thermophys. Conf., Snowmass 1980, AIAA Conf. Paper No.80-1486.

Wutz M.: 1965, Theorie und Praxis der Vakuumtechnik (Vieweg, Braunschweig).

Yuen W.W., Wong L.W.: 1979, A parametric study of radiative transfer with anisotropic scattering in a one-dimensional system, J. Quant. Spectrosc. Radiat. Transfer **22**, 231-238.

Yuen W.W., Tien C.L.: 1980, A successive approximation approach to problems in radiative transfer with a differential formulation, Trans. ASME, J. Heat Transfer **102**, 86-91.

Yuen W.W., Wong L.W.: 1980, Heat transfer by conduction and radiation in a one-dimensional absorbing, emitting and anisotropically scattering medium, Trans. ASME, J. Heat Transfer **102**, 303-307.

Yuen W.W., Wong L.W.: 1981, Effects of specular reflection on radiative transfer in an absorbing, emitting and anisotropically scattering medium, J. Quant. Spectrosc. Radiat. Transfer **25**, 427-434.

Yurevich F.B., Konyukh L.A.: 1975, Radiation attenuation by disperse media, Int. J. Heat Mass Transfer **18**, 819-829.

Ziegenbein B., Reiss H.: 1980, Thermische Isolierung, German patent appl. DE 30 46 032 A1 (filing date 6 Dec. 80), US patent appl. 4 425 413 (filing date 10 Jan. 84), Japanese patent appl. 81 P 480 (filing date 4 Dec. 81), French patent appl. 81 22728 (filing date 4 Dec. 81).

Ziegenbein B.: 1983, Evacuated high-temperature insulations for electrochemical batteries, Proc. 8th Eur. Conf. Thermophys. Prop., Baden-Baden 1982, High Temp.- High Pressures **15**, 329-334.

[illegible]

[illegible]

[illegible]

[illegible]

[illegible]

Subject Index